RSME Springer Series 13

As of 2015, RSME - Real Sociedad Matemática Española - and Springer cooperate in order to publish works by authors and volume editors under the auspices of a co-branded series of publications including advanced textbooks, Lecture Notes, collections of surveys resulting from international workshops and Summer Schools, SpringerBriefs, monographs as well as contributed volumes and conference proceedings. The works in the series are written in English only, aiming to offer high level research results in the fields of pure and applied mathematics to a global readership of students, researchers, professionals, and policymakers.

Henar Herrero · María Cruz Navarro ·
Hélia Serrano
Editors

Cutting-Edge Mathematics

RSME 2022, Ciudad Real, Spain, January 17–21, 2022

Real Sociedad Matemática
Española

Springer

Editors
Henar Herrero
Departamento de Matemáticas, Facultad de
Ciencias y Tecnologías Químicas (IMACI)
Universidad de Castilla-La Mancha
Ciudad Real, Spain

María Cruz Navarro
Departamento de Matemáticas, Facultad de
Ciencias y Tecnologías Químicas (IMACI)
Universidad de Castilla-La Mancha
Ciudad Real, Spain

Hélia Serrano
Departamento de Matemáticas, Facultad de
Ciencias y Tecnologías Químicas
Universidad de Castilla-La Mancha
Ciudad Real, Spain

ISSN 2509-8888 ISSN 2509-8896 (electronic)
RSME Springer Series
ISBN 978-3-031-62024-9 ISBN 978-3-031-62025-6 (eBook)
https://doi.org/10.1007/978-3-031-62025-6

This Springer imprint is published by the registered company Springer Nature Switzerland AG
The registered company address is: Gewerbestrasse 11, 6330 Cham, Switzerland

Contents

The Yang Problem for Complete Bounded Complex Submanifolds: a Survey

Antonio Alarcón(✉)

Departamento de Geometría y Topología e Instituto de Matemáticas (IMAG), Universidad de Granada, Granada, Spain

alarcon@ugr.es

Abstract. We survey the history as well as recent progress in the Yang problem concerning the existence of complete bounded complex submanifolds of the complex Euclidean spaces. We also point out some open questions on the topic.

1 The Problem and the First Solution

In 1977, P. Yang asked the following (see [66, p. 135, Question II]):

Problem 1. Do there exist complete immersed complex submanifolds $\varphi\colon M^k \to \mathbb{C}^n$ $(1 \leq k < n)$ with bounded image?

Completeness is a very natural condition to impose on a Riemannian manifold when one is interested in its global properties. Recall that an immersed submanifold $\varphi\colon M \to \mathbb{C}^n$ is said to be complete if the Riemannian metric $\varphi^* d\sigma^2$ induced on M by pulling back the Euclidean metric $d\sigma^2$ on $\mathbb{C}^n$ by the immersion φ is a complete metric on M: geodesics go on indefinitely. By the Hopf-Rinow theorem, φ is complete if and only if $\varphi \circ \gamma\colon [0, 1) \to \mathbb{C}^n$ has infinite Euclidean length for every divergent path $\gamma\colon [0, 1) \to M$. Every compact Riemannian manifold is complete, but compact complex manifolds cannot be found in $\mathbb{C}^n$ by the maximum principle for holomorphic functions, so the question in Problem 1 is in order.

The main original motivation for P. Yang to pose the aforementioned question is that a positive answer would prove the existence of complete immersed complex submanifolds $M^k \to \mathbb{C}^{2n}$ with strongly negative holomorphic sectional curvature [67]. On the other hand, since complex submanifolds of $\mathbb{C}^n$ are minimal (i.e., critical points for the volume functional; see e.g. [25,55], among many others, for an introduction to the subject), the Yang problem is also related to the so-called Calabi-Yau problem, which dates back to E. Calabi's conjectures from 1965 [54, p. 170] and asked whether there are complete bounded minimal hypersurfaces in $\mathbb{R}^n$ $(n \geq 3)$; we refer to [12, Ch. 7] for a recent survey on this fascinating topic. Nevertheless, the Yang problem for complex submanifolds has become an active focus of interest in its own right, having received many important contributions in the last decade; see [38, §4.3] for a brief introduction to the topic.

H. Herrero et al. (Eds.): RSME 2022, RSME 13, pp. 1–25, 2024.
https://doi.org/10.1007/978-3-031-62025-6_1

We shall denote by $\mathbb{D} = \{\zeta \in \mathbb{C} \colon |\zeta| < 1\}$ the open unit disc in $\mathbb{C}$ and by $\mathbb{B}_n = \{z = (z_1, \ldots, z_n) \in \mathbb{C}^n \colon |z| < 1\}$ the open unit ball in $\mathbb{C}^n$ for $n \geq 2$. The first affirmative answer to the question in Problem 1 was given only two years later, in 1979, by P.W. Jones. Recall that a map $f \colon X \to Y$ between topological spaces is said to be proper if $f^{-1}(K) \subset X$ is compact for every compact set $K \subset Y$.

Theorem 1 (Jones [51]). *There exist a complete bounded immersed complex disc $\mathbb{D} \to \mathbb{C}^2$, a complete bounded embedded complex disc $\mathbb{D} \hookrightarrow \mathbb{C}^3$, and a complete properly embedded complex disc $\mathbb{D} \hookrightarrow \mathbb{B}_4$.*

P.W. Jones' construction method is strongly complex analytic. It relies on using the BMO duality theorem in order to find a pair of bounded holomorphic functions f_1 and f_2 on $\mathbb{D}$ satisfying the property that

$$\int_\gamma \big(|f_1'(\zeta)| + |f_2'(\zeta)|\big)\, d\sigma(\zeta) = +\infty$$

for all paths $\gamma \subset \mathbb{D}$ terminating on $\mathbb{S}^1 = b\mathbb{D}$, where σ denotes Euclidean arc length. It follows that $\mathbb{D} \ni \zeta \mapsto (\zeta, f_1(\zeta), f_2(\zeta))$ is a complete bounded embedded complex disc in $\mathbb{C}^3$; the other two assertions in the theorem are obtained by slight modifications of this procedure.

2 Curves

Despite having to wait more than three decades for it, Theorem 1 has been generalized in several directions. The first extension of P.W. Jones' existence result was given by A. Alarcón and F.J. López in 2013 and concerns the topology of the examples.

Theorem 2 (Alarcón-López [15]). *Let $n \geq 2$. Every open orientable surface S admits a complex structure J such that the open Riemann surface $M = (S, J)$ carries a complete proper holomorphic immersion $M \to \mathbb{B}_n$ which is an embedding if $n \geq 3$. The same holds true if we replace the ball by any convex domain in $\mathbb{C}^n$.*

The embeddedness condition for $n \geq 3$ in this statement was not explicitly stated in [15]; nevertheless, it trivially follows from a standard transversality argument, as was later pointed out by A. Alarcón and F. Forstnerič in [6]. We emphasize that the examples in Theorem 2 may have any topological type, even infinite.

The proof in [15] is completely different from that in [51]. In particular, it is much more geometric, and is reminiscent of the method developed by N. Nadirashvili in his seminal paper [60] for constructing a complete bounded minimal disc in $\mathbb{R}^3$; see also [12, §7.1]. The construction goes by induction, the rough idea being the following. In the step $j \in \mathbb{N}$ we begin with a smoothly bounded compact complex curve, say X_{j-1}, whose boundary bX_{j-1} lies inside the ball

$R_{j-1}\mathbb{B}_n$ of some radius $R_{j-1} > 0$ but close to its boundary sphere. Then, we apply to X_{j-1} a deformation that is arbitrarily small outside a neighborhood of bX_{j-1} and pushes each boundary point $x \in bX_{j-1}$ a distance approximately $1/j$ in a direction perpendicular to the position vector of x in $\mathbb{C}^n$. In this way we increase the boundary distance from a fixed interior point of the curve an amount of approximately $1/j$, while the extrinsic diameter is increased, by Pythagoras' theorem, an amount of the order of $1/j^2$. Moreover, we ensure that the boundary of the new complex curve, X_j, lies inside the ball of radius

$$R_j = \sqrt{R_{j-1}^2 + 1/j^2} > R_{j-1}$$

but close to its boundary sphere. See Fig. 1. Since

$$\sum_{j\geq 1} 1/j = \infty \quad \text{and} \quad \sum_{j\geq 1} 1/j^2 < \infty,$$

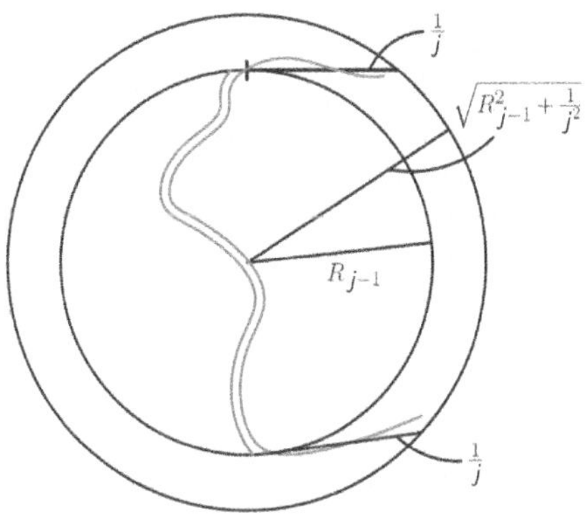

Fig. 1. Schematic representation of the geometry of the deformations used in the inductive construction in [15]. The key idea is to deform the submanifold near its boundary by pushing each boundary point in a direction orthogonal to its position vector.

if we arrange this process in the right way then we obtain in the limit a complete complex curve that is contained in the ball of radius $R = \lim_{j\to\infty} R_j < \infty$ and is proper in it. In order to prescribe the topology of the curve, we begin with a disc and at each step of the inductive construction apply a surgery which enables us to add either a handle or a boundary component to a given compact bordered complex curve in $\mathbb{C}^n$. Finally, if $n \geq 3$, then a general position argument allows to guarantee that all complex curves in the sequence are embedded, and hence the limit one can be granted to be embedded as well. The main tool in order to make the described deformations is the theory of uniform approximation for holomorphic functions on open Riemann surfaces; in particular, the Runge-Mergelyan theorem (see E. Bishop [21] or e.g. Theorem 5 in the survey on holomorphic approximation by J.E. Fornæss, F. Forstnerič, and E.F. Wold [33]).

A different construction of complete bounded complex curves in $\mathbb{C}^2$ can be found in [59], where F. Martín, M. Umehara, and K. Yamada gave examples with arbitrary finite genus and finitely many ends. Their technique relies on the existence of a simply connected complete bounded holomorphic null curve in $\mathbb{C}^3$ (such a curve was first constructed in [15], an alternative construction was given later by L. Ferrer, F. Martín, M. Umehara, and K. Yamada in [31]; see the mentioned sources or e.g. [12, §2.3] for definitions) and modifies a method developed by F.J. López in [56] for constructing complete minimal surfaces in $\mathbb{R}^3$ of hyperbolic conformal type.

Despite the flexibility of the methods in [15,59], none of them allows to control the complex structure on the curve, except of course in the simply connected case: every bounded immersed complex curve in $\mathbb{C}^n$ must have hyperbolic conformal type, and hence if it is simply connected it is biholomorphic to the disc $\mathbb{D}$. Indeed, since the construction in [15] relies on Runge's theorem, at a certain stage one does not have enough information on the placement in $\mathbb{C}^n$ of some parts of the curve, and hence one is forced to cut away some small pieces of the curve to ensure its boundedness. This makes one to lose the control of the complex structure of the curve. This difficulty was overcome by A. Alarcón and F. Forstnerič in [5], also published in 2013, where two additional complex analytic tools were introduced into the game, namely, the method of F. Forstnerič and E.F. Wold [43] for exposing boundary points of a complex curve in $\mathbb{C}^n$ and the use of approximate solutions to Riemann-Hilbert boundary value problems. (The former is a modern technique that has led to important progress in the classical Forster-Bell-Narasimhan Conjecture asking whether every open Riemann surface admits a proper holomorphic embedding into $\mathbb{C}^2$ [20,34]; we refer to F. Forstnerič [37, §9.10–9.11] for a survey on this long-standing open problem. On the other hand, the use of the Riemann-Hilbert problem for constructing proper holomorphic maps has a long history; we refer to F. Forstnerič and J. Globevnik [41], B. Drinovec Drnovšek and F. Forstnerič [28], and the references therein.) The implementation of these new tools enabled to substantially simplify the construction in [15] and, moreover, to control the complex structure on the curve. Recall that a bordered Riemann surface is an open connected Riemann surface M that is the interior, $M = \overline{M} \setminus b\overline{M}$, of a compact one dimensional complex manifold $\overline{M}$ with smooth boundary $b\overline{M}$ consisting of finitely many closed Jordan curves; such an $\overline{M}$ is called a compact bordered Riemann surface.

Theorem 3 (Alarcón-Forstnerič [5]). *Let $n \geq 2$ be an integer. Every bordered Riemann surface M admits a complete proper holomorphic immersion $M \to \mathbb{B}_n$ that is an embedding if $n \geq 3$. The same holds true if we replace the ball by any pseudoconvex domain in $\mathbb{C}^n$.*

More generally, it is shown in [5] that if X is a Stein manifold of dimension $n \geq 2$ endowed with a hermitian metric, then every bordered Riemann surface M admits a complete proper holomorphic immersion $M \to X$ that can be chosen an embedding if $n \geq 3$. Recall that a Stein manifold is the same thing as a closed complex submanifold of a complex Euclidean space; we refer to [37] for a comprehensive monograph on the theory of Stein manifolds. A domain D in $\mathbb{C}^n$

($n \geq 2$) is a Stein manifold if and only if it is pseudoconvex, meaning that there is a strictly plurisubharmonic exhaustion function $D \to \mathbb{R}$. This happens if and only if D is a domain of holomorphy and if and only if D is holomorphically convex. For instance, every convex domain in $\mathbb{C}^n$ is pseudoconvex. We refer to the monographs by R.M. Range [61] (see also the introductory note [62]) and L. Hörmander [49,50] for background on the subject.

More recently, the construction technique in [5] has been refined to produce complete bounded complex curves with control on the complex structure and with some further control on the asymptotic behavior. In particular, there are such curves bounded by Jordan curves. The following is a compilation of results by A. Alarcón, I. Castro Infantes, B. Drinovec Drnovšek, F. Forstnerič, and F.J. López [3,4,9,17,40].

Theorem 4. *Let R be a compact Riemann surface and assume that $M = R \setminus \bigcup_{i\in I} D_i$ is a domain in R whose complement is a finite or countable union of pairwise disjoint, smoothly bounded closed discs (i.e., diffeomorphic images of $\overline{\mathbb{D}}$). The following assertions hold true:*

1. *M is the complex structure of a complete bounded complex curve in $\mathbb{C}^2$. In fact, for any $n \geq 2$ there is a continuous map $\varphi\colon \overline{M} \to \mathbb{C}^n$ such that the restricted map $\varphi|_M\colon M \to \mathbb{C}^n$ is a complete holomorphic immersion (embedding if $n \geq 3$) and $\varphi|_{bM}\colon bM = \bigcup_{i\in I} bD_i \to \mathbb{C}^n$ is injective.* **(Alarcón-Drinovec Drnovšek-Forstnerič-López** [4], **Alarcón-Forstnerič** [9]. **)**
2. *If I is finite, then there is a continuous map $\varphi\colon \overline{M} \to \overline{\mathbb{B}}_n$ ($n \geq 3$) such that $\varphi(bM) \subset b\mathbb{B}_n$ and $\varphi|_M\colon M \to \mathbb{B}_n$ is a complete proper holomorphic immersion (embedding if $n \geq 3$). The same holds true with $\mathbb{B}_n$ replaced by any convex domain in $\mathbb{C}^n$.* **(Alarcón-Drinovec Drnovšek-Forstnerič-López** [4]. **)**
3. *If I is finite, then for any domain D in $\mathbb{C}^n$ there is a complete holomorphic immersion $\varphi\colon M \to D$ whose image is a dense subset of D. If $n \geq 3$ then φ can be chosen injective.* **(Alarcón-Castro Infantes** [3]. **)**
4. *There is a Cantor set C in R whose complement admits a complete holomorphic immersion $R \setminus C \to \mathbb{C}^2$ with bounded image. There also exist a Cantor set C in R and a complete holomorphic embedding $R \setminus C \hookrightarrow \mathbb{C}^3$ with bounded image.* **(Forstnerič** [40]. **)**
5. *There is a Cantor set C in M and a continuous map $\varphi\colon \overline{M} \setminus C \to \mathbb{C}^n$ ($n \geq 2$) such that $\varphi|_{M\setminus C}\colon M \setminus C \to \mathbb{C}^n$ is a complete holomorphic immersion and $\varphi|_{bM}\colon bM = \bigcup_{i\in I} bD_i \to \mathbb{C}^n$ is injective. If $n \geq 3$ then C can be chosen so that $\varphi\colon \overline{M} \setminus C \to \mathbb{C}^n$ is an injective map.* **(Forstnerič** [40]. **)**

Summarizing, by the year 2015 there were available in the literature several constructions of complete bounded complex curves immersed in $\mathbb{C}^2$ and embedded in $\mathbb{C}^3$, allowing a high control on the asymptotic behavior (proper in the ball or in a given pseudoconvex domain, bounded by Jordan curves, etc.), on the topology, and on the complex structure of the examples. However, the construction of complete bounded embedded complex curves in $\mathbb{C}^2$ turns out to be a much more challenging undertaking, and the question whether such curves exist remained open (see [5, Question 1]). Recall that complex curves are generically embedded in $\mathbb{C}^n$ for $n \geq 3$, meaning that self-intersections can be removed by applying small deformations, while self-intersections of complex curves in $\mathbb{C}^2$, which are generically double points, are stable under such deformations. That is the main reason why the task is a more difficult one.

The Yang problem for embedded complex curves in the affine plane $\mathbb{C}^2$ was finally settled by A. Alarcón and F.J. López in a paper published in 2016.

Theorem 5 (Alarcón-López [18]**).** *Every convex domain in $\mathbb{C}^2$ admits complete properly embedded complex curves.*

The proof goes by induction and involves an approximation process by embedded complex curves in $\mathbb{C}^2$. The main step in the construction is to prove that every compact embedded complex curve X in $\mathbb{C}^2$ with the boundary bX lying in the boundary bD of a regular strictly convex domain $D \subset \mathbb{C}^2$ may be approximated by another compact embedded complex curve $\tilde{X}$ with $b\tilde{X} \subset b\tilde{D}$, for any given convex domain $\tilde{D} \subset \mathbb{C}^2$ with $\overline{D} \subset \tilde{D}$. The new curve $\tilde{X}$ is ensured to contain a biholomorphic copy of X, which we denote by X as well, and the main point is to guarantee that $\tilde{X} \setminus X \subset \tilde{D} \setminus D$ and the intrinsic Euclidean distance in $\tilde{X}$ from X to $b\tilde{X}$ is suitably larger (in a Pythagorical way similar to that explained in Fig. 1) than the distance from D to $b\tilde{D}$. These conditions are the key for obtaining embeddedness, completeness, and properness of the limit complex curve in the limit convex domain. In order to guarantee the embeddedness of $\tilde{X}$ a standard self-intersection removal method consisting of replacing every normal crossing in an immersed complex curve in $\mathbb{C}^2$ by an embedded annulus is applied. This surgery may generate shortcuts in the arising desingularized curve $\tilde{X}$, thereby giving rise to divergent paths of shorter length. This is an important difficulty for ensuring completeness; for instance, if one applies this surgery at each step in the inductive construction in the proof of Theorem 2 or Theorem 3, then one still obtains in the limit a properly embedded complex curve in $\mathbb{B}_2$ (or, more generally, in any given pseudoconvex domain of $\mathbb{C}^2$), but it need not be complete. A main novelty in the construction in [18] is provided a good enough (say, in a Pythagorical sense) estimate of the growth of the intrinsic Euclidean diameter of the desingularized, embedded complex curve $\tilde{X}$; this is achieved by keeping a stronger control on the placement of $\tilde{X} \setminus X$ in $\tilde{D} \setminus D$. The main tool in this construction continues to be the classical Runge-Mergelyan approximation theorem for holomorphic functions on open Riemann surfaces.

The aforementioned surgery may increase the topological genus of the curve, and so there is no control on the topology of the examples in Theorem 5. They actually seem to have infinite genus, and can be ensured to have infinite topology. It therefore remained an open question whether there are complete bounded embedded complex curves in $\mathbb{C}^2$ of finite topology (see [18, Question 1.5]).

3 Submanifolds of Arbitrary Dimension

All examples of complete bounded complex submanifolds we have discussed so far are of complex dimension one; i.e., complex curves in $\mathbb{C}^n$. The first known such submanifolds of higher dimension were also given in [5].

Corollary 1 (Alarcón-Forstnerič [5]). *If D is a relatively compact, strongly pseudoconvex domain in a Stein manifold of dimension $k \geq 1$, then there are a complete proper holomorphic immersion $D \to (\mathbb{B}_2)^{2k} \subset \mathbb{C}^{4k}$ and a complete proper holomorphic embedding $D \hookrightarrow (\mathbb{B}_3)^{2k+1} \subset \mathbb{C}^{6k+3}$.*

The corollary is obtained by the following simple trick which was pointed out by J.E. Fornæss. In [29], B. Drinovec Drnovšek and F. Forstnerič proved that every domain D as in the statement admits a proper holomorphic immersion $g\colon D \to \mathbb{D}^{2k}$ into the polydisc $\mathbb{D}^{2k} = \mathbb{D}\times \overset{2k}{\cdots} \times\mathbb{D} \subset \mathbb{C}^{2k}$. Choose a complete proper holomorphic immersion $\varphi\colon \mathbb{D} \to \mathbb{B}_2$ provided by Theorem 3, and consider the proper holomorphic map $\varphi^{2k}\colon \mathbb{D}^{2k} \to (\mathbb{B}_2)^{2k}$ given by

$$\varphi^{2k}(\zeta_1,\ldots,\zeta_{2k}) = (\varphi(\zeta_1),\ldots,\varphi(\zeta_{2k})), \quad (\zeta_1,\ldots,\zeta_{2k}) \in \mathbb{D}^{2k}.$$

It is then easily checked that $\varphi^{2k}\circ g\colon D \to (\mathbb{B}_2)^{2k}$ is a complete proper holomorphic immersion. A slight modification of this argument using a proper holomorphic embedding $D \hookrightarrow \mathbb{D}^{2k+1}$ (existence of such is also proved in [29]) provides a proper holomorphic embedding $D \hookrightarrow (\mathbb{B}_3)^{2k+1}$. This same trick together with Theorem 5 allows to prove the following.

Corollary 2 (Alarcón-López [18]). *For any $k \in \mathbb{N}$ there is a complete bounded embedded k-dimensional complex submanifold $M^k \hookrightarrow \mathbb{C}^{2k}$.*

The ad hoc construction of the high dimensional examples in Corollaries 1 and 2 seemed to give very particular solutions to the Yang problem for complex submanifolds of dimension ≥ 2. This led to some new questions, as whether the dimension $2k$ in Corollary 2 is optimal, and exposed the need of looking for new construction methods other than those based on the existence of complete bounded complex curves. For instance, the natural question whether the ball $\mathbb{B}_k$ ($k \geq 2$) admits a complete proper holomorphic embedding into the ball $\mathbb{B}_n$ for some $n > k$, and, in particular, for $n = k+1$ appeared; see [5, Question 3] and [44, Question 13.2]. An affirmative answer to this question in the case of sufficiently high codimension was given in 2015 by B. Drinovec Drnovšek.

Theorem 6 (Drinovec Drnovšek [27]). *Every bounded strictly pseudoconvex domain $D \subset \mathbb{C}^k$ ($k \in \mathbb{N}$) with $\mathcal{C}^2$ boundary admits a complete proper holomorphic embedding $D \hookrightarrow \mathbb{B}_n$ for any large enough $n \in \mathbb{N}$.*

The proof in [27] continues to use the geometric idea of deforming a compact submanifold near the boundary in orthogonal directions to the position vector (see Fig. 1), but it exploits different tools. The main new ingredients are holomorphic peak functions, going back to ideas of M. Hakim and N. Sibony [48] and E. Løw [57], and the construction of inner functions on the ball, as well as J.E.

Fornæss embedding theorem [32]. The construction relies on suitably modifying earlier methods by F. Forstnerič [35] and E. Løw [58] for constructing proper holomorphic maps from a strictly convex domain with $\mathcal{C}^2$ boundary in $\mathbb{C}^k$ into a unit ball of some Euclidean space of higher dimension, in order to make them complete. The construction method in [27] requires a sufficiently high codimension, and hence the following question remains open; see [5, Question 3] and [44, Question 13.2].

Problem 2. Does there exist a complete proper holomorphic embedding $\mathbb{B}_k \hookrightarrow \mathbb{B}_{k+1}$ for $k \geq 2$?

4 Hypersurfaces

Recall that complex submanifolds of dimension $k \in \mathbb{N}$ in $\mathbb{C}^n$ are generically embedded for $n \geq 2k+1$ (meaning that one can get rid of their self-intersections by applying small deformations on compact pieces), while for $n \leq 2k$ self-intersections of k-dimensional complex submanifolds in $\mathbb{C}^n$ are stable under small deformations. This is the main reason why the Yang problem for embedded submanifolds is much more difficult in low codimension; in particular, for hypersurfaces. In the lowest dimensional case of $n = 2$ (i.e., for complex curves in $\mathbb{C}^2$), this question was first solved by A. Alarcón and F.J. López in [18] (see Theorem 5 above), but the question whether there are complete bounded embedded complex hypersurfaces in $\mathbb{C}^n$, or even whether there are such complex submanifolds of dimension k with $2k \geq n$, remained open for every $n \geq 3$. Indeed, note that the codimension in all examples of complete bounded embedded complex submanifold of dimension ≥ 2 which have been mentioned so far is high.

It was J. Globevnik who, in a pair of landmark papers in 2015–2016, positively settled Yang's question in Problem 1 for embeddings in arbitrary dimension and codimension; in particular, for hypersurfaces.

Theorem 7 (Globevnik [44,46]). *For any pair of integers $1 \leq k < n$ there is a complete closed embedded k-dimensional complex submanifold of $\mathbb{B}_n$. In particular, $\mathbb{B}_n$ admits a complete properly embedded complex hypersurface.*

The same holds true if we replace the ball by any pseudoconvex domain in $\mathbb{C}^n$.

Globevnik's approach is completely different from any previous method used in the study of the Yang problem, and it was a major breakthrough in this topic. In particular, his construction of a complete closed complex hypersurface in a given pseudoconvex domain $D \subset \mathbb{C}^n$ is implicit: the examples are obtained as level sets of highly oscillating holomorphic functions on D. (The existence of complete closed complex submanifolds of any higher codimension is then an obvious consequence.) To be more precise, Globevnik proved the following.

Theorem 8 (Globevnik [44,46]). *For any pseudoconvex domain $D \subset \mathbb{C}^n$ ($n \geq 2$) there is a holomorphic function on D whose real part is unbounded above on every divergent path $\gamma\colon [0, 1) \to D$ of finite length.*

If a function $f\colon D \to \mathbb{C}$ on a domain $D \subset \mathbb{C}^n$ is holomorphic and nonconstant, then all its nonempty level sets $f^{-1}(c) = \{z \in D\colon f(z) = c\}$ ($c \in \mathbb{C}$) are closed complex hypersurfaces of D, possibly with singularities. Nevertheless, by Sard's theorem most of them are smooth, and hence properly embedded. So, if $c \in \mathbb{C}$ is such that $f^{-1}(c) \neq \varnothing$, then every divergent path $\gamma\colon [0,1) \to f^{-1}(c)$ diverges on D as well. If D is pseudoconvex and the function f is as in Theorem 8 then, since $\Re(f)$ is constant (and hence bounded) on γ, we have that γ has infinite length, and $f^{-1}(c)$ is thus complete. Therefore, the level sets $f^{-1}(c)$ ($c \in \mathbb{C}$) of f form a (possibly singular) holomorphic foliation of D by complete closed complex hypersurfaces. Theorem 8 thus implies the following corollary which, in turn, implies Theorem 7.

Corollary 3 (Globevnik [44,46]). *Every pseudoconvex domain $D \subset \mathbb{C}^n$ ($n \geq 2$) admits a (possibly singular) holomorphic foliation by complete closed complex hypersurfaces (most of which are smooth).*

Let us outline the proof of Theorem 8 in the case of $D = \mathbb{B}_n$ given in [44]. Recall that a convex polytope P in $\mathbb{R}^d$, $d \geq 2$, is a compact convex set which is the intersection of finitely many closed half-spaces. A face of P is a closed convex subset $F \subset P$ such that every closed segment in P whose relative interior intersects F is contained in F. The boundary bP of P is the union of its faces of dimension $d-1$, while the skeleton skel(P) of P is the union of all $(d-2)$-dimensional faces of P. Most of the work in the proof consists of constructing a sequence of convex polytopes P_j in $\mathbb{C}^n = \mathbb{R}^{2n}$ and positive numbers θ_j ($j \in \mathbb{N}$) satisfying the following conditions:

(i) $P_1 \subset \mathrm{Int}(P_2) \subset P_2 \subset \mathrm{Int}(P_3) \subset \cdots \subset \bigcup_{j\in\mathbb{N}} P_j = \mathbb{B}_n$.
(ii) Denote by U_j the θ_j-neighborhood of skel(P_j) in bP_j, and set $V_j = (bP_j)\setminus U_j$, $j \in \mathbb{N}$. If $\gamma\colon [0,1) \to \mathbb{B}_n$ is a divergent path such that $\gamma([0,1)) \cap V_j = \varnothing$ for all $j \geq j_0$ for some $j_0 \in \mathbb{N}$, then γ has infinite length.

Each set V_j ($j \in \mathbb{N}$) is compact and its connected components are closed convex sets in hyperplanes of $\mathbb{R}^{2n}$. The union

$$L = \bigcup_{j\in\mathbb{N}} V_j \tag{1}$$

of all of them is a sort of labyrinth of compact connected $(2n-1)$-dimensional convex sets in $\mathbb{B}_n \subset \mathbb{R}^{2n}$ with the property that every divergent path $\gamma\colon [0,1) \to \mathbb{B}_n$ meeting at most finitely many components of L has infinite length. See Fig. 2.

The construction of the labyrinth in [44] is very involved and belongs to convex geometry. With the labyrinth in hand, to complete the proof of Theorem 8 an idea of J. Globevnik and E.L. Stout from [47] is used in order to construct, via Runge's theorem, a sequence of holomorphic polynomials $f_j\colon \mathbb{C}^n \to \mathbb{C}$, $j \in \mathbb{N}$, such that the following conditions hold for each $j \in \mathbb{N}$:

(iii) $\Re(f_j(z)) \geq j+1$ for all $z \in V_j$.
(iv) $|f_{j+1}(z) - f_j(z)| \leq 1/2^{j+1}$ for all $z \in P_j$.

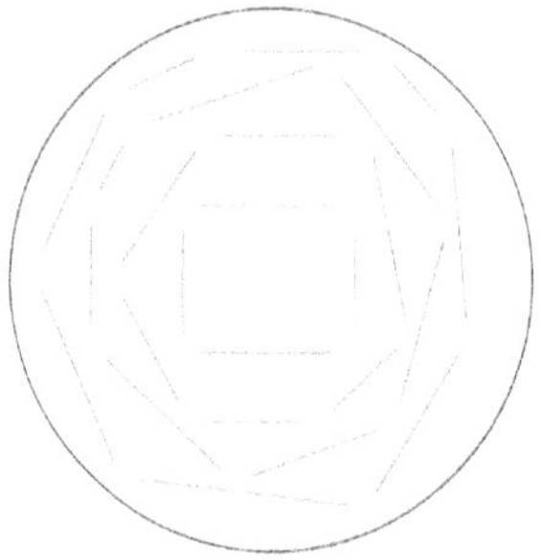

Fig. 2. Globevnik's labyrinth. Each piece of the labyrinth is a compact convex set in a real affine hyperplane. Every divergent path in the ball meeting at most finitely many components of the labyrinth has infinite length.

Conditions (i) and (iv) ensure that there is a limit holomorphic function

$$f = \lim_{j\to\infty} f_j \colon \mathbb{B}_n \to \mathbb{C}.$$

By conditions (iii) and (iv), we have $|f(z) - f_j(z)| < 1$ and hence $\Re(f(z)) \geq j$ for all $z \in V_j$, $j \in \mathbb{N}$. Since every divergent path $\gamma\colon [0,1) \to \mathbb{B}_n$ of finite length meets V_j for infinitely many j's by condition (ii), we obtain that $\Re(f)$ is unbounded above on every such γ, proving Theorem 8 in the case of $D = \mathbb{B}_n$.

The proof of the theorem for an arbitrary pseudoconvex domain in $\mathbb{C}^n$ $(n \geq 2)$ given in [46] relies on a modification of this technique, exploiting the well known fact that every such domain is biholomorphic to a properly embedded complex submanifold of a complex Euclidean space; see [49, Theorem 5.3.9]. (In fact, if $n \geq 2$ then every n-dimensional Stein manifold admits a proper holomorphic embedding into $\mathbb{C}^N$ for $N = [\frac{3n}{2}] + 1$; see Y. Eliashberg and M. Gromov [30] for even n and J. Schürmann [63] for odd $n \geq 3$. The dimension $[\frac{3n}{2}] + 1$ in this assertion is the lowest prossible for all $n \geq 2$ due to purely topological reasons, as O. Forster pointed out by examples in [34].)

Theorem 8 also holds for $n = 1$ the proof being much easier. In fact, it is not hard to show, as an application of the classical Runge theorem in one variable, that every domain $\Omega \subset \mathbb{C}$ admits a holomorphic function whose real part is unbounded both above and below on every divergent path $[0,1) \to \Omega$ of finite length (recall that every domain in $\mathbb{C}$ is pseudoconvex). In this direction, P.W. Jones proved back in 1980 that the unit disc $\mathbb{D}$ admits a bounded holomorphic function h such that all level sets of $|h|\colon \mathbb{D} \to [0,\infty)$ have infinite length [52]. Obviously, the functions in Theorem 8 are not bounded, hence the following remains an open (but likely difficult) question.

Problem 3. Does the open unit ball $\mathbb{B}_n$ $(n \geq 2)$ admit a bounded holomorphic function whose level sets are all complete?

4.1 Controlling the Topology of the Examples

Neither the construction in [18] of complete properly embedded complex curves in the ball $\mathbb{B}_2$ nor that in [44, 46] of complete properly embedded complex hypersurfaces in $\mathbb{B}_n$ for arbitrary $n \geq 2$ provide any control on the topology of the examples, which, in principle, could be very complicated. In particular, it remained an open question whether there are complete properly embedded complex curves of finite topology in the ball of $\mathbb{C}^2$, and, more precisely, whether there are simply connected ones; see [18, Question 1.5] and [44, Question 13.1].

The first construction of complete properly embedded complex hypersurfaces in the ball with control of the topology was carried out shortly after by A. Alarcón, J. Globevnik, and F.J. López, who proved the following.

Theorem 9 (Alarcón-Globevnik-López [14]). *Assume that $Z \subset \mathbb{C}^n$ $(n \geq 2)$ is a smooth closed complex hypersurface with $Z \cap \mathbb{B}_n \neq \varnothing$ and let $K \subset Z \cap \mathbb{B}_n$ be a connected compact set. Then there exist a pseudoconvex Runge domain $X \subset Z$ such that $K \subset X$ and a complete proper holomorphic embedding $X \hookrightarrow \mathbb{B}_n$.*

In particular, $\mathbb{B}_n$ admits smooth complete closed complex hypersurfaces which are biholomorphic to a pseudoconvex Runge domain in $\mathbb{C}^{n-1}$.

In the case of $n = 2$, Theorem 9 trivially implies the following more precise result. Recall that all open orientable surfaces can be realized as Runge domains of properly embedded complex curves in $\mathbb{C}^2$; see M. Černe and F. Forstnerič [22] for the finite topological case and A. Alarcón and F.J. López [16] for the arbitrary one.

Corollary 4 (Alarcón-Globevnik-López [14]). *The open unit ball $\mathbb{B}_2$ of $\mathbb{C}^2$ admits complete properly embedded complex curves with any finite topology.*

In particular, there is a complete proper holomorphic embedding $\mathbb{D} \hookrightarrow \mathbb{B}_2$.

This settled in the positive the aforementioned questions [18, Question 1.5] and [44, Question 13.1]. The construction method in [14] is different from the ones in [18, 44, 46]. In particular, it introduces a new tool into the game: the use of holomorphic automorphisms of $\mathbb{C}^n$. In order to prove Theorem 9 one begins with the given properly embedded hypersurface Z in $\mathbb{C}^n$ and the natural embedding $Z \hookrightarrow \mathbb{C}^n$ given by the inclusion map. Then, in a recursive way, one composes this initial embedding with a sequence of holomorphic automorphisms Φ_j of $\mathbb{C}^n$, $j \in \mathbb{N}$, which converges uniformly on compact subsets of $\mathbb{B}_n$. One obtains in this way a sequence of proper holomorphic embeddings $\Psi_j \colon Z \hookrightarrow \mathbb{C}^n$, where $\Psi_j = \Phi_j \circ \Phi_{j-1} \circ \cdots \circ \Phi_1$, $j \in \mathbb{N}$, whose images converge uniformly on compact subsets of $\mathbb{B}_n$ to a properly embedded complex hypersurface of $\mathbb{B}_n$. If this process is carried out in the right way, then one can guarantee that a connected component V of the limit hypersurface is biholomorphic to a pseudoconvex Runge domain X in Z which is closed in $\mathbb{B}_n$ and contains K. The main point is that the sequence of holomorphic automorphisms Φ_j of $\mathbb{C}^n$ can be chosen to make sure that V is disjoint from a labyrinth L of compact sets in $\mathbb{B}_n$ as that constructed by J. Globevnik in [44] (see (1) and Fig. 2), and hence V is in addition complete. Indeed, denote by $\{L_j\}_{j \in \mathbb{N}}$ the family of connected components of L and note that, up to reordering, the following properties are satisfied for all $j \in \mathbb{N}$:

(a) L_j is a compact convex set contained in an affine real hyperplane H_j of $\mathbb{C}^n$.
(b) There exists a compact convex set $K_j \subset \mathbb{C}^n$ such that $\bigcup_{i=1}^{j-1} L_i \subset K_j$ and $K_j \cap H_l = \varnothing$ for all $l \geq j$.
Moreover, up to removing finitely many pieces of the labyrinth if necessary, we may also assume that:
(c) There is a compact convex set $K_0 \subset \mathbb{C}^n$ such that $K \subset K_0 \subset \mathbb{C}^n \setminus \bigcup_{l\in\mathbb{N}} H_l$.

By Kallin's lemma (see E. Kallin [53] or E.L. Stout [65, p. 62]) and the Oka-Weil theorem (see [65, Theorem 1.5.1]), if C and T are disjoint compact convex sets in $\mathbb{C}^n$ and $A \subset C$ is a compact polynomially convex set, then the union $A \cup T$ is polynomially convex as well; we refer to [65] for a monograph on polynomial convexity. This easily implies in view of conditions (a), (b), and (c) that:

(d) The set $K_j \cup L_j$ is polynomially convex for all $j \in \mathbb{N} \cup \{0\}$. More generally, the compact set $K_j \cup \bigcup_{i=j}^{k} L_k$ is polynomially convex for all $k \geq j$, $j \in \mathbb{N} \cup \{0\}$.

In this situation, the Andersén-Lempert theory (see E. Andersén and L. Lempert [19] and F. Forstnerič and J.-P. Rosay [42]) enables one to construct a sequence $\{\Phi_j\}_{j\in\mathbb{N}}$ of holomorphic automorphisms of $\mathbb{C}^n$ such that if

$$\Psi_j = \Phi_j \circ \cdots \circ \Phi_1, \quad j \in \mathbb{N},$$

then $|\Phi_j(z) - z| < \epsilon_j$ for all $z \in K_j$ and

$$\Psi_j(Z) \cap \bigcup_{i=1}^{j} L_j = \varnothing \quad \text{for all } j \in \mathbb{N}.$$

Here, each ϵ_j is a positive number which may be chosen arbitrarily small at each step in the inductive construction. In fact, setting $\Psi_0 = \Phi_0 = \mathrm{Id}_{\mathbb{C}^n}$ and assuming that we have such automorphisms $\Phi_1, \ldots, \Phi_{j-1}$ for some $j \in \mathbb{N}$, up to a suitable affine complex change of coordinates we can even choose Φ_j to be a shear map of the form

$$\Phi_j(\zeta, w) = (\zeta, e^{\phi_j(\zeta)} w), \quad \zeta \in \mathbb{C},\ w \in \mathbb{C}^{n-1},$$

with a holomorphic function $\phi_j : \mathbb{C} \to \mathbb{C}$. Choosing the number $\epsilon_j > 0$ sufficiently small at each step, we have that $\lim_{j\to\infty} \Psi_j = \Psi$ exists uniformly on compact subsets in $\Omega := \bigcup_{j\in\mathbb{N}} \Psi_j^{-1}(K_j)$, and $\Psi\colon \Omega \to \mathbb{B}_n$ is a biholomorphic map such that

$$\Psi(Z \cap \Omega) \cap L = \varnothing.$$

This implies that $X = Z \cap \Omega$ is a pseudoconvex Runge domain in Z, $K \subset X$, and the closed complex hypersurface $V = \Psi(X)$ of $\mathbb{B}_n$ is complete (since it is disjoint from the labyrinth L in (1); see condition (ii) in the description of L), thereby proving Theorem 9.

Despite the fact that the method developed in [14] works fine if one uses the labyrinths constructed in [44], the authors introduced in [14] new labyrinths of compact sets in the ball with properties analogous to those in (i)-(iv) and (a)-(d) above. The main reason for building these new labyrinths is that they are constructed in a much simpler way; only basic trigonometry is required.

Definition 1. *Given an integer* $n \geq 2$ *and a pair of positive numbers* $0 < r < R$*, we say that a compact set* $\mathcal{L}$ *in the spherical shell* $R\mathbb{B}_n \setminus r\overline{\mathbb{B}}_n = \{z \in \mathbb{C}^n : r < |z| < R\}$ *is a tangent labyrinth if the following conditions are satisfied:*

- $\mathcal{L}$ *has finitely many connected components* $T_1, \ldots, T_k$ $(k \in \mathbb{N})$.
- *Each component* T_j *of* $\mathcal{L}$ *is a closed round ball in a real affine hyperplane in* $\mathbb{C}^n$ *which is orthogonal to the position vector of the center* x_j *of the ball* T_j.
- *If* $|x_i| = |x_j|$ *for some* $i, j \in \{1, \ldots, k\}$*, then the radii of the balls* T_i *and* T_j *are equal, while if* $|x_i| < |x_j|$ *then* $T_i \subset |x_j|\mathbb{B}_n$.

These new labyrinths are much more symmetric than Globevnik's ones in [44], and its pieces are all round balls. See Fig. 3.

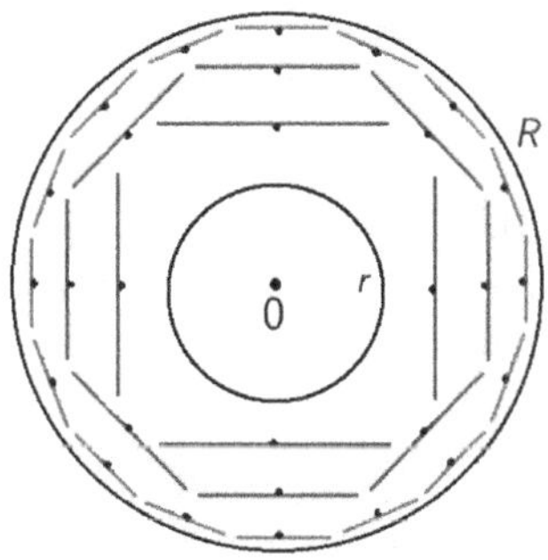

Fig. 3. A tangent labyrinth in the spherical shell $R\mathbb{B}_n \setminus r\overline{\mathbb{B}}_n$. Each piece of the labyrinth is a closed round ball in a real affine hyperplane. The pieces of the labyrinth are organized in layers according to the norm of the center of each piece.

The main two properties of tangent labyrinths are the following:

(I) If $\mathcal{L}$ is a tangent labyrinth in the spherical shell $R\mathbb{B}_n \setminus r\overline{\mathbb{B}}_n$ ($0 < r < R$, $n \geq 2$), then the compact set $r\overline{\mathbb{B}}_n \cup \mathcal{L} \subset \mathbb{C}^n$ is polynomially convex.

(II) For any given numbers $0 < r < R$, $\delta > 0$ (big), and $\tau > 0$ (small) there exists a tangent labyrinth $\mathcal{L}$ in $R\mathbb{B}_n \setminus r\overline{\mathbb{B}}_n$ ($n \geq 2$) enjoying the following conditions:
- If $\gamma\colon [0,1] \to \mathbb{B}_n$ is a path crossing the shell (i.e., $|\gamma(0)| \leq r$ and $|\gamma(1)| \geq R$) and γ is disjoint from the labyrinth (i.e., $\gamma([0,1]) \cap \mathcal{L} = \varnothing$), then the Euclidean length of γ is greater than δ.
- Each component T of $\mathcal{L}$ has Euclidean diameter smaller than τ; i.e. the radius of the Euclidean ball T is smaller than $\tau/2$.

Condition (I) can be seen as a consequence of the Kallin lemma and the Oka-Weil theorem, as above; this was observed by A. Alarcón and F. Forstnerič in [7], see also [1, Remark 2.4]. On the other hand, condition (II) was granted in [14, Lemma 2.4]; see also [1, Lemma 2.3].

Note that the tangent labyrinths $\mathcal{L}$ in Definition 1 are compact and have finitely many components, while those L in (1) are closed in the ball and have infinitely many components. It is however obvious in view of condition (II) that for any increasing sequence $0 < r_1 < R_1 < r_2 < R_2 < \cdots$ with $\lim_{j\to\infty} r_j = 1$ there is a sequence of tangent labyrinths $\mathcal{L}_j$ in $R_j\mathbb{B}_n \setminus r_j\overline{\mathbb{B}}_n$, $j \in \mathbb{N}$, with the property that every divergent path $\gamma\colon [0,1) \to \mathbb{B}_n$ meeting at most finitely many components of the infinite labyrinth

$$\mathcal{L} = \bigcup_{j\in\mathbb{N}} \mathcal{L}_j \subset \mathbb{B}_n \tag{2}$$

has infinite length; cf. property (ii) of the Globevnik labyrinth L in (1).

The new construction method developed in [14], based on the use of tangent labyrinths and holomorphic automorphisms of $\mathbb{C}^n$, has given rise to some further developments. For instance, A. Alarcón and J. Globevnik extended Corollary 4 to complex curves of arbitrary topology, possibly infinite, including in addition a hitting condition.

Theorem 10 (Alarcón-Globevnik [13]). *Given a closed discrete subset Λ of $\mathbb{B}_2$ and an open orientable smooth surface M, there is a complex structure J on M such that the open Riemann surface (M,J) admits a complete proper holomorphic embedding $(M,J) \hookrightarrow \mathbb{B}_2$ whose image contains Λ.*

Choosing the set Λ such that the limit set $\overline{\Lambda} \setminus \Lambda$ equals the boundary sphere $b\mathbb{B}_n = \{z \in \mathbb{C}^n\colon |z| = 1\}$, one obtains the following immediate corollary.

Corollary 5 (Alarcón-Globevnik [13]). *$\mathbb{B}_2$ contains complete properly embedded complex curves with any topology whose limit set is the whole sphere $b\mathbb{B}_2$.*

In the case when M is simply-connected, and hence the Riemann surface (M,J) is biholomorphic to the disc $\mathbb{D}$, Theorem 10 was previously obtained by J. Globevnik in [45]. The proof of Theorem 10 combines ideas from [14] and [16,43], the latter relying on the method of exposing boundary points of bordered Riemann surfaces in $\mathbb{C}^2$ as well as on the use of Fatou-Bieberbach domains of $\mathbb{C}^2$. Theorem 10 is somehow analogous to the main result in [16] where properly embedded complex curves in $\mathbb{C}^2$ with arbitrary topology are constructed.

Also using tangent labyrinths and holomorphic automorphisms of $\mathbb{C}^n$ as in [14], A. Alarcón and F. Forstnerič obtained complete complex submanifolds in $\mathbb{B}_n$ with controlled topology, having no self-intersections, and hitting any given countable subset (not necessarily closed or discrete), thereby complementing Theorem 9.

Theorem 11 (Alarcón-Forstnerič [7]). *Let $\Lambda \subset \mathbb{B}_n$ ($n \geq 2$) be a countable set. If Z is a smooth closed complex submanifold of $\mathbb{C}^n$ and $K \neq \varnothing$ is a connected compact subset of $Z \cap \mathbb{B}_n$, then there are a pseudoconvex Runge domain $X \subset Z$ such that $K \subset X$ and a complete holomorphic injective immersion $\varphi\colon X \hookrightarrow \mathbb{B}_n$ with $\Lambda \subset \varphi(X)$. In particular, $\varphi(X)$ can be made dense in $\mathbb{B}_n$.*

In the case of complex curves, this has the following immediate corollary; compare with Theorem 10 and statement 3 in Theorem 4.

Corollary 6 (Alarcón-Forstnerič [7]). *Given a countable subset* Λ *of* $\mathbb{B}_n$ $(n \geq 2)$ *and an open orientable smooth surface* M *of finite topology, there is a complex structure* J *on* M *such that the open Riemann surface* (M, J) *admits a complete holomorphic injective immersion* $\varphi\colon (M, J) \hookrightarrow \mathbb{B}_n$ *whose image contains* Λ.

In particular, $\varphi(M)$ *can be made dense in* $\mathbb{B}_n$.

The first result which provided control of the topology of all leaves in a holomorphic foliation of the ball by complete closed complex submanifolds was subsequently given by A. Alarcón and F. Forstnerič.

Theorem 12 (Alarcón-Forstnerič [8]). *There exists a nonsingular holomorphic foliation of the ball* $\mathbb{B}_n$ $(n \geq 2)$ *all of whose leaves are complete properly embedded complex discs in* $\mathbb{B}_n$.

The topology of the leaves in the foliations given in Theorem 12 is the simplest possible one: discs. It is not known whether a comparable result holds for leaves with prescribed but more complicated topology. The proof follows a similar approach as that in [14], but with a more careful use of the Andersén-Lempert theory. The idea is, using holomorphic automorphisms, to successively twist a holomorphic foliation of $\mathbb{C}^n$ by complex lines in order to make bigger and bigger parts of the foliation avoid more and more pieces of an infinite labyrinth in $\mathbb{B}_n$, as those in (1) or (2). Each component of the intersection of the ball with a properly embedded complex line is Runge in the complex line, and hence simply connected. Since it is in addition bounded, it is a properly embedded complex disc in the ball. More precisely, the construction provides a sequence of foliations $\{\mathcal{F}_j\}_{j\in\mathbb{N}}$ of $\mathbb{B}_n$ by properly embedded complex discs, converging to a limit foliation $\mathcal{F}$ whose all leaves are discs and each one meets at most finitely many components of the labyrinth, so it is complete.

None of the methods discussed in this section enables one to control the complex structure of the examples, except of course if the curve is simply connected. In particular, the following remains an open question; see [13, Problem 1.5].

Problem 4. Let $\overline{M} = M \cup bM$ be a compact bordered Riemann surface admitting a smooth embedding $\overline{M} \hookrightarrow \mathbb{C}^2$ which is holomorphic on M. Does there exist a complete holomorphic embedding $M \hookrightarrow \mathbb{C}^2$ with bounded image?

F. Forstnerič and E.F. Wold proved in [43] that every bordered Riemann surface M as in Problem 4 admits a proper holomorphic embedding $M \hookrightarrow \mathbb{C}^2$. Obviously, such an embedding is complete but unbounded. Here is a less ambitious but still appealing question in this direction.

Problem 5. Let $\overline{M}$ be as in Problem 4. Does there exist a complete non-proper holomorphic embedding $M \hookrightarrow \mathbb{C}^2$?

4.2 Complete Complex Hypersurfaces in the Ball Come in Foliations

In principle, it was not clear whether the construction of Globevnik in [44] gives, say, a rare type or a very special type of solution to the embedded Yang problem for hypersurfaces in the ball. Note that his examples of complete closed complex hypersurfaces in $\mathbb{B}_n$ are obtained as leaves of a holomorphic foliation of $\mathbb{B}_n$ by complex hypersurfaces of the same sort. Thus, the question whether every such hypersurface can be embedded into a foliation of the ball by hypersurfaces with the same properties naturally appeared. Recall that, at that time, there were already available in the literature several constructions of complete properly embedded complex submanifolds in balls, besides the implicit one, so one could be inclined to think that the answer to this question is negative. However, the answer is positive, as it follows from the following more general result by A. Alarcón dealing with submanifolds of arbitrary codimension.

Theorem 13 (Alarcón [1]). *Let n and q be integers with $1 \leq q < n$, let V be a smooth closed complex submanifold of pure codimension q in $\mathbb{B}_n$, possibly disconnected, and assume that V is contained in a fibre of a holomorphic submersion $\mathbb{B}_n \to \mathbb{C}^q$. Then there is a holomorphic submersion $f\colon \mathbb{B}_n \to \mathbb{C}^q$ such that:*

- *V is a union of components of $f^{-1}(0)$.*
- *The fibre $f^{-1}(c) \subset \mathbb{B}_n$ is a smooth complete closed complex submanifold of pure codimension q for every $c \in f(\mathbb{B}_n) \setminus \{0\}$.*
- *$f^{-1}(0) \setminus V$ is either empty or a smooth complete closed complex submanifold of pure codimension q in $\mathbb{B}_n$.*
- *If V is a fibre of a holomorphic submersion $\mathbb{B}_n \to \mathbb{C}^q$, then f can be chosen with $f^{-1}(0) = V$.*

The family of components of the fibres $f^{-1}(c)$ ($c \in \mathbb{C}^q$) of the holomorphic submersion f given by the theorem form a nonsingular holomorphic submersion foliation of $\mathbb{B}_n$ by smooth connected closed complex submanifolds of codimension q all which, except perhaps those contained in the initial submanifold $V \subset f^{-1}(0)$, are complete. If the given V is complete, then all the leaves in the foliation are complete. This provided the first known example of a foliation of the ball $\mathbb{B}_n$ by complete connected closed complex submanifolds of codimension $q > 1$, as well as the first example of a nonsingular holomorphic foliation of $\mathbb{B}_n$ by such submanifolds for any dimension $n \geq 2$ and codimension $1 \leq q < n$.

Somewhat surprisingly, the hypersurface V in Theorem 13 need not be complete, so the following holds.

Corollary 7 (Alarcón [1,2]). *Let n and q be a pair of integers with $1 \leq q < n$. The following assertions hold:*

1. *There exists a nonsingular holomorphic submersion foliation of $\mathbb{B}_n$ by smooth connected complete closed complex submanifolds of codimension q.*

2. *For any integer $k \geq 1$, there exists a nonsingular holomorphic submersion foliation of $\mathbb{B}_n$ by smooth connected closed complex submanifolds of codimension q all which, except precisely k among them, are complete.*
3. *There exists a nonsingular holomorphic submersion foliation of $\mathbb{B}_n$ by smooth connected closed complex submanifolds of codimension q all which, except precisely countably infinitely many, are complete.*

The following result concerning the existence of foliations of the ball having both many complete leaves and many incomplete leaves was subsequently obtained by a generalization of the construction method in [1].

Corollary 8 (Alarcón [2]). *Given integers n and q with $1 \leq q < n$, there is a nonsingular holomorphic submersion foliation $\mathcal{F}$ of the ball $\mathbb{B}_n$ by smooth connected closed complex submanifolds of codimension q such that both the union of the complete leaves of $\mathcal{F}$ and the union of the incomplete leaves of $\mathcal{F}$ are dense subsets of $\mathbb{B}_n$. In particular, every leaf of $\mathcal{F}$ is both a limit leaf of the family of complete leaves of $\mathcal{F}$ and a limit leaf of the family of incomplete leaves of $\mathcal{F}$.*

The assumption in Theorem 13 that the given hypersurface V is contained in a fibre of a holomorphic submersion is necessary and cannot be relaxed. In the case of $q = 1$, this condition is always satisfied (see F. Forstnerič [39]), so the theorem applies to every smooth closed complex hypersurface in $\mathbb{B}_n$. More precisely, it is known that every closed complex hypersurface V (possibly with singularities) in $\mathbb{B}_n$ is defined by a holomorphic function on $\mathbb{B}_n$ which is noncritical out of the singular set of V. (This actually holds true with the ball replaced by any Stein manifold X with the vanishing second cohomology group $H^2(X; Z) = 0$ [39]; which extends the classical result by J.-P. Serre from 1953 [64] that every divisor in such a Stein manifold X is a principal divisor.) The following more precise version of Theorem 13 in the case of hypersurfaces, possibly with singularities, was also obtained in [1].

Theorem 14 (Alarcón [1]). *For any closed complex hypersurface V (possibly with singularities) in $\mathbb{B}_n$ $(n \geq 2)$ there exists a holomorphic function f on $\mathbb{B}_n$ satisfying the following conditions:*

- $V = f^{-1}(0)$.
- *The critical locus of f coincides with the singular set of V.*
- *The level set $f^{-1}(c) \subset \mathbb{B}_n$ is a smooth complete closed complex hypersurface for every $c \in f(\mathbb{B}_n) \setminus \{0\}$.*

In this case, the family of components of the level sets $f^{-1}(c)$ $(c \in \mathbb{C})$ of the holomorphic function f furnished by the theorem form a holomorphic foliation of $\mathbb{B}_n$ by connected closed complex hypersurfaces all which, except perhaps those contained in the given hypersurface $V = f^{-1}(0)$, are smooth and complete. If the given V is smooth, then f is noncritical and hence the foliation nonsingular; while if V is complete, then all the leaves in the foliation are complete. This established a sort of converse to Theorem 8, namely, it proves that, in fact, every complete closed complex hypersurface in $\mathbb{B}_n$ is defined by a holomorphic function on $\mathbb{B}_n$ whose level sets are all complete (i.e., as those constructed by Globevnik in [44]), and hence it can be embedded as a leaf in a holomorphic foliation of the ball by complete closed complex hypersurfaces. Moreover, the defining function can be chosen noncritical, and hence the foliation nonsingular, whenever the given hypersurface is smooth.

The method used in the proof of Theorem 13 broadly follows the one developed by Globevnik for proving Theorem 8 in [44], but it presents some major differences and novelties. The holomorphic submersion $f\colon \mathbb{B}_n \to \mathbb{C}^q$ satisfying the conclusion of the theorem is obtained as the limit of a sequence of holomorphic submersions $f_j\colon \mathbb{B}_n \to \mathbb{C}^q$ ($j \in \mathbb{N}$), and it is granted that each fibre of f, except perhaps the one over $0 \in \mathbb{C}^q$, meets at most finitely many components of a suitable infinite labyrinth of compact sets in $\mathbb{B}_n$, thereby making sure that every such fibre is complete. A main novelty is that the proof requires to construct the labyrinth and the limit holomorphic submersion f at the same time in an inductive procedure; that is, the labyrinth cannot be fixed beforehand, as in the proofs in [14,44] and in the aforementioned subsequent papers. Moreover, the labyrinth depends on the given submanifold V, and the components of the labyrinth that meet V are treated in a completely different way than those that do not; in particular, the former components must be chosen with small diameter. Another novelty of the proof is that it relies on a rather sophisticated Oka-Weil-Cartan type approximation with interpolation theorem for holomorphic submersions due to F. Forstnerič (see [36,39] or [37, §9.12–9.16]), instead of the more basic Runge-Weierstrass theorem. We next spell out some of the details.

Let us briefly outline the proof of Theorem 13 in the slightly simpler case when there is a holomorphic submersion $f_0\colon \mathbb{B}_n \to \mathbb{C}^q$ such that $V = f_0^{-1}(0)$; that is always the case if $q = 1$. Choose any increasing sequence of positive numbers $0 < r_1 < R_1 < r_2 < R_2 < \cdots \to 1$, fix $\epsilon_0 > 0$, choose $0 < R_0 < r_1$, and call $L_0 = \varnothing$. The proof consists of inductively constructing a sequence $S_j = \{f_j, \epsilon_j, \mathcal{L}_j\}_{j\in\mathbb{N}}$, where

- $f_j\colon \mathbb{B}_n \to \mathbb{C}^q$ is a holomorphic submersion,
- $\epsilon_j > 0$ is a positive number, and
- $\mathcal{L}_j$ is a tangent labyrinth in the spherical shell $R_j\mathbb{B}_n \setminus r_j\overline{\mathbb{B}}_n$ (see Definition 1),

such that the following conditions are satisfied for each $j \in \mathbb{N}$:

(1_j) $|f_j(z) - f_{j-1}(z)| < \epsilon_j$ for all $z \in r_j\overline{\mathbb{B}}_n$.

(2_j) $f_j^{-1}(0) = V$.

(3_j) $0 < 2\epsilon_j < \epsilon_{j-1}$ and if $f\colon \mathbb{B}_n \to \mathbb{C}^q$ is a holomorphic map such that $|f(z) - f_{j-1}(z)| < 2\epsilon_j$ for all $z \in r_j\overline{\mathbb{B}}_n$, then f is submersive everywhere on $R_{j-1}\mathbb{B}_n$.

(4_j) If $\gamma\colon [0,1] \to \mathbb{B}_n$ is a path satisfying that $|\gamma(0)| \le r_j$, $|\gamma(1)| \ge R_j$, and $\gamma([0,1]) \cap \mathcal{L}_j = \varnothing$, then $\operatorname{length}(\gamma) > 1$.

(5_j) If $z \in \mathcal{L}_j$, then either $|f_j(z)| > j$ or $|f_j(z)| < 1/j$.

Carrying out this process in the right way, one obtains a holomorphic submersion

$$f = \lim_{j\to\infty} f_j \colon \mathbb{B}_n \to \mathbb{C}^q, \quad j \in \mathbb{N},$$

as well as an infinite labyrinth of compact sets

$$\mathcal{L} = \bigcup_{j\in\mathbb{N}} \mathcal{L}_j \subset \mathbb{B}_n,$$

satisfying the following conditions:

(A) $f^{-1}(0) = V$.

(B) Every divergent path $[0,1) \to \mathbb{B}_n$ hitting at most finitely many components of $\mathcal{L}$ has infinite length.

(C) $f^{-1}(c)$ meets at most finitely many components of $\mathcal{L}$ for each $c \in \mathbb{C}^q \setminus \{0\}$.

Note that conditions (B) and (C) are granted by properties (4_j) and (5_j), respectively. It is clear that (B) and (C) ensure that $f^{-1}(c)$ is a smooth complete closed complex submanifold of pure codimension q in $\mathbb{B}_n$ for all $c \in \mathbb{C}^q \setminus \{0\}$, and hence, in view of condition (A), f satisfies the conclusion of the theorem.

For the inductive step, fix $j \in \mathbb{N}$ and assume that we already have a suitable triple $S_{j-1} = \{f_{j-1}, \epsilon_{j-1}, \mathcal{L}_{j-1}\}$. First of all, choose a number $\epsilon_j > 0$ so small that (3_j) holds; that is always possible by the Cauchy estimates. Also choose, as we may by (2_{j-1}), a number $\tau > 0$ so small that

$$|f_{j-1}(z)| < 1/j \quad \text{for all } z \in R_j\overline{\mathbb{B}}_n \text{ with } \operatorname{dist}(z, V) < \tau. \tag{3}$$

Next, take a tangent labyrinth $\mathcal{L}_j$ in $R_j\mathbb{B}_n \setminus r_j\overline{\mathbb{B}}_n$ satisfying (4_j) and such that

$$\operatorname{diam}(T) < \tau \quad \text{for each component } T \text{ of } \mathcal{L}_j. \tag{4}$$

(See property (II) below Definition 1.) We then split the labyrinth $\mathcal{L}_j$ into two disjoint parts, $\mathcal{L}_j = \Lambda_V \cup \Lambda_0$, where Λ_V is the union of those components of $\mathcal{L}_j$ which intersect V and Λ_0 is the union of those which are disjoint from V. Both Λ_V and Λ_0 consist of finitely many components and $\Lambda_V \cap \Lambda_0 = \varnothing$. So, we have in view of (3) and (4) that

$$|f_{j-1}(z)| < 1/j \quad \text{for all } z \in \Lambda_V. \tag{5}$$

In this situation, a standard application of the aforementioned Oka-Weil-Cartan-Forstnerič theorem for holomorphic submersions (see [1, Theorem 2.1] for the precise statement that we need here), taking into account that $r_j\overline{\mathbb{B}}_n \cup \mathcal{L}_j$ is a polynomially convex compact set (see property (I) below Definition 1), furnishes us with a holomorphic submersion $f_j \colon \mathbb{B}_n \to \mathbb{C}^q$ satisfying the following conditions:

- $|f_j(z) - f_{j-1}(z)| < \delta$ for all $z \in r_j\overline{\mathbb{B}}_n \cup \Lambda_V$ for any given $\delta > 0$.

- $f_j^{-1}(0) = V$.
- $|f_j(z)| > j$ for all $z \in \Lambda_0$.

It is clear that these properties, together with (5) and the fact that $\mathcal{L}_j = \Lambda_V \cup \Lambda_0$, guarantee that f_j satisfies conditions (1_j), (2_j), and (5_j) provided that the number $\delta > 0$ is chosen sufficiently small. This closes the induction and completes the sketch of the proof of Theorem 13.

Complete complex hypersurfaces in a Stein manifold equipped with a Riemannian metric come in foliations as well, as the following result points out.

Theorem 15 (Alarcón [1]). *Let X be a Stein manifold of dimension ≥ 2 equipped with a Riemannian metric, and assume that V is a smooth closed complex hypersurface in X such that the normal bundle $N_{V/X}$ of V in X is trivial. Then there exists a holomorphic function f on X satisfying the following conditions:*

- *$V \subset f^{-1}(0)$.*
- *The level set $f^{-1}(c)$ is complete for every $c \in f(X) \setminus \{0\}$.*
- *$f^{-1}(0) \setminus V$ is either empty or a complete closed complex hypersurface (possibly with singularities) in X.*
- *f in noncritical everywhere on V.*

Completeness in this statement refers to the metric induced in the hypersurface by that in X. Note that the assumption on the normal bundle $N_{V/X}$ is necessary by the F. Docquier and H. Grauert tubular neighborhood theorem [26]. The connected components of the level sets $f^{-1}(c) \subset X$ ($c \in \mathbb{C}$) of the holomorphic function f provided by Theorem 15 form a (possibly singular) holomorphic foliation of the Stein manifold X by connected closed complex hypersurfaces all which, except perhaps those contained in V, are complete. If V is complete, then all the leaves are complete. Moreover, most level sets of f are smooth by Sard's theorem, and hence we obtain the following extension of Corollary 3. (Recall that pseudoconvex domains are Stein manifolds, but not every Stein manifold is biholomorphic to a pseudoconvex domain in a complex Euclidean space.)

Corollary 9 (Alarcón [1]). *Every Stein manifold X of dimension ≥ 2 equipped with a Riemannian metric admits a (possibly singular) holomorphic foliation by complete connected closed complex hypersurfaces (possibly with singularities).*

In particular, X admits a complete properly embedded complex hypersurface.

The proof of Theorem 15 combines the construction method in [1] with a slight refinement of the trick used by Globevnik in [46] to show that every pseudoconvex domain (other than the ball) in $\mathbb{C}^n$ admits a holomorphic foliation by complete closed complex hypersurfaces (see Theorem 8).

Although every Stein manifold admits a holomorphic function without critical points, as was shown by F. Forstnerič in [36], the method of proof in [1] does not allow to guarantee that the function f furnished by Theorem 15 be noncritical. Therefore, the following remains an open problem (see [1, Conjecture 5.2]).

Problem 6 Does every Stein manifold X of dimension $n \geq 2$ equipped with a Riemannian metric admit a nonsingular holomorphic foliation by smooth complete closed complex hypersurfaces?

4.3 Labyrinths in Pseudoconvex Domains

Recently, S. Charpentier and Ł. Kosiński [23] constructed labyrinths of compact sets in any given pseudoconvex Runge domain $D \subset \mathbb{C}^n$ with properties analogous to those of the tangent labyrinths in balls (see Definition 1 and properties (I) and (II) below it).

Theorem 16 (Charpentier-Kosiński [23]). *Let $\Omega \subset \mathbb{C}^n$ $(n \geq 2)$ be a smooth strongly pseudoconvex domain and let $K \subset \Omega$ be a $\mathscr{O}(\Omega)$-convex compact set. Then, for any $\delta > 0$ there is a compact set $\Gamma \subset \Omega \setminus K$ satisfying the following conditions:*

- *Γ has finitely many components all of which are holomorphically contractible.*
- *$K \cup \Gamma$ is $\mathscr{O}(\Omega)$-convex.*
- *If $\gamma\colon [0,1) \to \Omega$ is a divergent path with $\gamma(0) \in K$ and $\gamma([0,1)) \cap \Gamma = \varnothing$, then* $\mathrm{length}(\gamma) > \delta$.
- *If Ω is a Runge domain, then the components of Γ can be chosen to be images of Euclidean balls of real dimension $2n-1$ under $\mathbb{R}$-affine isomorphisms and automorphisms of $\mathbb{C}^n$.*

Since every pseudoconvex domain $D \subset \mathbb{C}^n$ admits a normal exhaustion $\Omega_1 \Subset \Omega_2 \Subset \cdots \subset \bigcup_{j\in\mathbb{N}} \Omega_j = D$ by smooth $\mathscr{O}(D)$-convex strongly pseudoconvex domains, which can be taken to be Runge if D is Runge (see [37, §2.3]), an inductive application of Theorem 16 provides holomorphically contractible compact sets $\Gamma_j \subset \Omega_{j+1} \setminus \overline{\Omega}_j$, $j \in \mathbb{N}$, satisfying the following conditions (see [23, Theorem 1.1]):

- Γ_j has finitely many components and $\overline{\Omega}_j \cup \Gamma_j$ is $\mathscr{O}(D)$-convex for all $j \in \mathbb{B}_n$.
- If $\gamma\colon [0,1) \to D$ is a divergent path meeting at most finitely many components of $\Gamma = \bigcup_{j\in\mathbb{N}} \Gamma_j$, then γ has infinite length.
- If Ω is a Runge domain, then the components of Γ can be chosen to be images of Euclidean balls of real dimension $2n-1$ under $\mathbb{R}$-affine isomorphisms and automorphisms of $\mathbb{C}^n$.

This can be used to extend from the ball $\mathbb{B}_n$ to any pseudoconvex Runge domain D in $\mathbb{C}^n$ many of the results mentioned in this survey, just by using the labyrinths in Theorem 16 instead of the tangent labyrinths in Definition 1. In particular, all results in Sect. 4.1, as well as Theorem 13, Theorem 14, and the first assertion in Corollary 7, hold true with the ball replaced by any pseudoconvex Runge domain (for the extension of Theorem 12 it is also required that the domain D be hyperbolic, meaning that it does not contain any biholomorphic copy of $\mathbb{C}$). Note, in particular, that this provides an affirmative answer to Problem 6 in the special case when the Stein manifold X is a pseudoconvex Runge domain in $\mathbb{C}^n$. Likewise, Corollary 8 and assertions 2 and 3 in Corollary 7 hold true with the ball replaced by any such domain D if we assume in addition that D contains no smooth closed complex submanifold of $\mathbb{C}^n$; this happens for instance if $D \subset \mathbb{C}^n$ is relatively compact.

In the subsequent paper [24], S. Charpentier and Ł. Kosiński extended Globevnik's existence result for highly oscillating holomorphic functions in Theorem 8, to the existence of functions with an even wilder asymptotic behavior.

Theorem 17 (Charpentier-Kosiński [24]). *Let $D \subset \mathbb{C}^n$ $(n \geq 2)$ be a pseudoconvex domain. Then there are holomorphic functions f on D with the property that $f(\gamma([0,1)))$ is everywhere dense in $\mathbb{C}$ for any divergent path $\gamma\colon [0,1) \to D$ with finite length. In fact, the set of all such functions is a residual, densely lineable, spaceable subset of the space of all holomorphic functions on D endowed with the locally uniform convergence topology.*

Note that the functions f provided by Theorem 17 also satisfy the conclusion of Theorem 8. In particular, the components of the level sets $f^{-1}(c)$ $(c \in \mathbb{C})$ form a (possibly singular) holomorphic foliation of D by complete connected closed complex hypersurfaces (possibly with singularities), as those in Corollary 3. We emphasize that the functions f in Theorem 17 are abundant; in particular, they form a dense subset of the space of all holomorphic functions on D. Therefore, the functions in Theorem 8 are abundant as well.

Added in proof

After this paper was written, Problems 5 and 6 were affirmatively settled by Alarcón and Forstnerič in [10] and [11], respectively.

Acknowledgments. This research was partially supported by the State Research Agency (AEI) via the grant no. PID2020-117868GB-I00 and the "Maria de Maeztu" Excellence Unit IMAG, reference CEX2020-001105-M, funded by MCIN/AEI/10.13039/501100011033/; and the Junta de Andalucía grant no. P18-FR-4049; Spain.

I wish to thank Franc Forstnerič for some remarks that led to improve the exposition.

References

1. Alarcón, A.: Complete complex hypersurfaces in the ball come in foliations. J. Differential Geom. **121**(1), 1–29 (2022)
2. Alarcón, A.: Wild holomorphic foliations of the ball. Indiana Univ. Math. J. **71**(2), 561–578 (2022)
3. Alarcón, A., Castro-Infantes, I.: Complete minimal surfaces densely lying in arbitrary domains of $\mathbb{R}^n$. Geom. Topol. **22**(1), 571–590 (2018)

4. Alarcón, A., Drinovec Drnovšek, B., Forstnerič, F., López, F.J.: Every bordered Riemann surface is a complete conformal minimal surface bounded by Jordan curves. Proc. Lond. Math. Soc. (3), **111**(4), 851–886 (2015)
5. Alarcón, A., Forstnerič, F.: Every bordered Riemann surface is a complete proper curve in a ball. Math. Ann. **357**(3), 1049–1070 (2013)
6. Alarcón, A., Forstnerič, F.: Null curves and directed immersions of open Riemann surfaces. Invent. Math. **196**(3), 733–771 (2014)
7. Alarcón, A., Forstnerič, F.: Complete densely embedded complex lines in $\mathbb{C}^2$. Proc. Amer. Math. Soc. **146**(3), 1059–1067 (2018)
8. Alarcón, A., Forstnerič, F.: A foliation of the ball by complete holomorphic discs. Math. Z. **296**(1–2), 169–174 (2020)
9. Alarcón, A., Forstnerič, F.: The Calabi-Yau problem for Riemann surfaces with finite genus and countably many ends. Rev. Mat. Iberoam. **37**(4), 1399–1412 (2021)
10. Alarcón, A., Forstnerič, F.: Embedded complex curves in the affine plane. Ann. Mat. Pura Appl. **203**(4), 1673–1701 (2024)
11. Alarcón, A., Forstnerič, F.: Complete nonsingular holomorphic foliations on Stein manifolds. Mediterr. J. Math. **21**(1), 25 (2024)
12. Alarcón, A., Forstnerič, F., López, F.J.: Minimal Surfaces from a Complex Analytic Viewpoint. Springer Monographs in Mathematics, Springer, Cham (2021). https://doi.org/10.1007/978-3-030-69056-4
13. Alarcón, A., Globevnik, J.: Complete embedded complex curves in the ball of $\mathbb{C}^2$ can have any topology. Anal. PDE **10**(8), 1987–1999 (2017)
14. Alarcón, A., Globevnik, J., López, F.J.: A construction of complete complex hypersurfaces in the ball with control on the topology. J. Reine Angew. Math. **751**, 289–308 (2019)
15. Alarcón, A., López, F.J.: Null curves in $\mathbb{C}^3$ and Calabi-Yau conjectures. Math. Ann. **355**(2), 429–455 (2013)
16. Alarcón, A., López, F.J.: Proper holomorphic embeddings of Riemann surfaces with arbitrary topology into $\mathbb{C}^2$. J. Geom. Anal. **23**(4), 1794–1805 (2013)
17. Alarcón, A., López, F.J.: Compact complete null curves in complex 3-space. Israel J. Math. **195**(1), 97–122 (2013)
18. Alarcón, A., López, F.J.: Complete bounded embedded complex curves in $\mathbb{C}^2$. J. Eur. Math. Soc. (JEMS) **18**(8), 1675–1705 (2016)
19. Andersén, E., Lempert, L.: On the group of holomorphic automorphisms of C^n. Invent. Math. **110**(2), 371–388 (1992)
20. Bell, S.R., Narasimhan, R.: Proper holomorphic mappings of complex spaces. In: Several complex variables, VI, Encyclopaedia Math. Sci., pp. 1–38. Springer, Berlin (1990). https://doi.org/10.1007/978-3-642-61299-2_1
21. Bishop, E.: Subalgebras of functions on a Riemann surface. Pacific J. Math. **8**, 29–50 (1958)
22. Černe, M., Forstnerič, F.: Embedding some bordered Riemann surfaces in the affine plane. Math. Res. Lett. **9**(5–6), 683–696 (2002)
23. Charpentier, S., Kosiński, Ł: Construction of labyrinths in pseudoconvex domains. Math. Z. **296**(3–4), 1021–1025 (2020)
24. Charpentier, S., Kosiński, Ł: Wild boundary behaviour of holomorphic functions in domains of $\mathbb{C}^N$. Indiana Univ. Math. J. **70**(6), 2351–2367 (2021)
25. Colding, T.H., Minicozzi, W.P., II.: Minimal submanifolds. Bull. London Math. Soc. **38**(3), 353–395 (2006)
26. Docquier, F., Grauert, H.: Levisches Problem und Rungescher Satz für Teilgebiete Steinscher Mannigfaltigkeiten. Math. Ann. **140**, 94–123 (1960)

27. Drinovec Drnovšek, B.: Complete proper holomorphic embeddings of strictly pseudoconvex domains into balls. J. Math. Anal. Appl. **431**(2), 705–713 (2015)
28. Drinovec Drnovšek, B., Forstnerič, F.: Holomorphic curves in complex spaces. Duke Math. J. **139**(2), 203–253 (2007)
29. Drinovec Drnovšek, B., Forstnerič, F.: Strongly pseudoconvex domains as subvarieties of complex manifolds. Amer. J. Math. **132**(2), 331–360 (2010)
30. Eliashberg, Y., Gromov, M.: Embeddings of Stein manifolds of dimension n into the affine space of dimension $3n/2+1$. Ann. of Math. (2), **136**(1), 123–135 (1992)
31. Ferrer, L., Martín, F., Umehara, M., Yamada, K.: A construction of a complete bounded null curve in C^3. Kodai Math. J. **37**(1), 59–96 (2014)
32. Fornaess, J.E.: Embedding strictly pseudoconvex domains in convex domains. Amer. J. Math. **98**(2), 529–569 (1976)
33. Fornæss, J.E., Forstnerič, F., Wold, E.F.: Holomorphic approximation: the legacy of Weierstrass, Runge, Oka–Weil, and Mergelyan. In: Breaz, D., Rassias, M.T. (eds.) Advancements in Complex Analysis, pp. 133–192. Springer, Cham (2020). https://doi.org/10.1007/978-3-030-40120-7_5
34. Forster, O.: Plongements des variétés de Stein. Comment. Math. Helv. **45**, 170–184 (1970)
35. Forstnerič, F.: Embedding strictly pseudoconvex domains into balls. Trans. Amer. Math. Soc. **295**(1), 347–368 (1986)
36. Forstnerič, F.: Noncritical holomorphic functions on Stein manifolds. Acta Math. **191**(2), 143–189 (2003)
37. Forstnerič, F.: Stein manifolds and holomorphic mappings (The homotopy principle in complex analysis). Ergebnisse der Mathematik und ihrer Grenzgebiete. 3. Folge,, vol. 56, 2nd edn. Springer, Cham (2017). https://doi.org/10.1007/978-3-319-61058-0
38. Forstnerič, F.: Holomorphic embeddings and immersions of Stein manifolds: a survey. In: Byun, J., Cho, H., Kim, S., Lee, KH., Park, JD. (eds.) Geometric Complex Analysis. Springer Proceedings in Mathematics & Statistics, vol. 246, pp. 145–169. Springer, Singapore (2018). https://doi.org/10.1007/978-981-13-1672-2_11
39. Forstnerič, F.: Divisors defined by noncritical functions. Proc. Amer. Math. Soc. **146**(7), 2985–2994 (2018)
40. Forstnerič, F.: The Calabi-Yau problem for minimal surfaces with Cantor ends. Rev. Mat. Iberoam. **39**(6), 2067–2077 (2023)
41. Forstnerič, F., Globevnik, J.: Discs in pseudoconvex domains. Comment. Math. Helv. **67**(1), 129–145 (1992)
42. Forstnerič, F., Rosay, J.-P.: Approximation of biholomorphic mappings by automorphisms of C^n. Invent. Math. **112**(2), 323–349 (1993)
43. Forstnerič, F., Wold, E.F.: Bordered Riemann surfaces in $\mathbb{C}^2$. J. Math. Pures Appl. (9), **91**(1), 100–114 (2009)
44. Globevnik, J.: A complete complex hypersurface in the ball of $\mathbb{C}^N$. Ann. of Math. (2), **182**(3), 1067–1091 (2015)
45. Globevnik, J.: Embedding complete holomorphic discs through discrete sets. J. Math. Anal. Appl. **444**(2), 827–838 (2016)
46. Globevnik, J.: Holomorphic functions unbounded on curves of finite length. Math. Ann. **364**(3–4), 1343–1359 (2016)
47. Globevnik, J., Stout, E.L.: Holomorphic functions with highly noncontinuable boundary behavior. J. Analyse Math. **41**, 211–216 (1982)
48. Hakim, M., Sibony, N.: Fonctions holomorphes bornées sur la boule unité de C^n. Invent. Math. **67**(2), 213–222 (1982)

49. Hörmander, L.: An Introduction to Complex Analysis in Several Variables.North-Holland Mathematical Library, vol. 7, 3rd edn. North-Holland Publishing Co., Amsterdam (1990)
50. Hörmander, L.: Notions of Convexity. Progress in Mathematics, vol. 127. Birkhäuser Boston, Inc., Boston (1994)
51. Jones, P.W.: A complete bounded complex submanifold of C^3. Proc. Amer. Math. Soc. **76**(2), 305–306 (1979)
52. Jones, P.W.: Bounded holomorphic functions with all level sets of infinite length. Michigan Math. J. **27**(1), 75–79 (1980)
53. E. Kallin. Polynomial convexity: The three spheres problem. In: Aeppli, A., Calabi, E., Röhrl, H. (eds.) Proceedings of the Conference on Complex Analysis, pp. 301–304. Springer, Heidelberg (1965). https://doi.org/10.1007/978-3-642-48016-4_26
54. Kobayashi, S., Eells Jr., J.: Proceedings of the United States-Japan Seminar in Differential Geometry, Kyoto, Japan, 1965. Nippon Hyoronsha Co., Ltd., Tokyo (1966)
55. Lawson Jr., H.B.: Lectures on minimal submanifolds. Vol. I. Mathematics Lecture Series, vol. 9, 2nd edn. Publish or Perish, Inc., Wilmington (1980)
56. López, F.J.: Hyperbolic complete minimal surfaces with arbitrary topology. Trans. Amer. Math. Soc. **350**(5), 1977–1990 (1998)
57. Løw, E.: A construction of inner functions on the unit ball in C^p. Invent. Math. **67**(2), 223–229 (1982)
58. Løw, E.: Embeddings and proper holomorphic maps of strictly pseudoconvex domains into polydiscs and balls. Math. Z. **190**(3), 401–410 (1985)
59. Martín, F., Umehara, M., Yamada, K.: Complete bounded holomorphic curves immersed in $\mathbb{C}^2$ with arbitrary genus. Proc. Amer. Math. Soc. **137**(10), 3437–3450 (2009)
60. Nadirashvili, N.: Hadamard's and Calabi-Yau's conjectures on negatively curved and minimal surfaces. Invent. Math. **126**(3), 457–465 (1996)
61. Range, R.M.: Holomorphic Functions and Integral Representations in Several Complex Variables. Graduate Texts in Mathematics, vol. 108. Springer-Verlag, New York (1986). https://doi.org/10.1007/978-1-4757-1918-5
62. Range, R.M.: What is... a pseudoconvex domain? Notices Amer. Math. Soc. **59**(2), 301–303 (2012)
63. Schürmann, J.: Embeddings of Stein spaces into affine spaces of minimal dimension. Math. Ann. **307**(3), 381–399 (1997)
64. Serre, J.-P.: Quelques problèmes globaux relatifs aux variétés de Stein. In *Colloque sur les fonctions de plusieurs variables, tenu à Bruxelles, 1953*, pp. 57–68. Georges Thone, Liège; Masson & Cie, Paris (1953)
65. Stout, E.L.: Polynomial Convexity. Progress in Mathematics, vol. 261. Birkhäuser Boston, Inc., Boston (2007)
66. Yang, P.: Curvature of complex submanifolds of C^n. In: Several complex variables (Proc. Sympos. Pure Math., Vol. XXX, Part 2, Williams Coll., Williamstown, Mass., 1975), pp. 135–137. Amer. Math. Soc., Providence, R.I. (1977)
67. Yang, P.: Curvatures of complex submanifolds of $\mathbf{C}^n$. J. Differential Geom., **12**(4), 499–511 (1978). 1977

Phylogenetic Reconstruction Based on Algebra

Marta Casanellas(✉)

Universitat Politècnica de Catalunya, Diagonal 628, Barcelona, Spain
marta.casanellas@upc.edu

Abstract. Phylogenetics is the discipline that studies the evolutionary history of species. During the last years there has been an approach to phylogenetics from the point of view of algebraic geometry. This perspective has been used to study the evolutionary models most used in phylogenetics but also to develop new phylogenetic reconstruction tools. Here we review the interplay between algebra and phylogenetics and we explain the most recent results that use these methods for the purpose of phylogenetic reconstruction.

1 Introduction

According to Darwin's theory of natural selection, the evolution of species is usually represented on a *phylogenetic tree*: its leaves represent living species, the interior nodes represent their common ancestors, and edges represent evolutionary processes (see Fig. 1). Nowadays, the study of the evolutionary history of a group of species is carried out from deoxyribonucleic acid (DNA) molecules associated to the living species in the study. These DNA molecules may correspond to certain genes or to other molecular entities and, due to the double helix structure of DNA, they can be seen as words on the alphabet $\{\mathtt{A}, \mathtt{C}, \mathtt{G}, \mathtt{T}\}$ with $\mathtt{A}$ denoting adenine, $\mathtt{C}$ cytosine, $\mathtt{G}$ guanine, and $\mathtt{T}$ thymine. In this sense they are called *DNA sequences*. The aim of phylogenetics is to reconstruct the ancestral relationships among species (i.e. the phylogenetic tree) from a given set of DNA sequences.

Phylogenetics is not only aimed at the knowledge of evolutionary history per se: it has applications on many different topics nowadays. For example, as pointed out in [3], phylogenetics is used in the design of biodiversity preservation policies, in the prediction of molecular evolution of viruses, or in the detection of tumor origins. It has recently been used in determining the origin of SARS-CoV-2 and the phylogenetic tree of coronaviruses has been crucial in detecting the relationship between this coronavirus in bats and humans, see [4]. Phylogenetics has also an impact beyond biology: it is a crucial tool in the study of origin and evolution of languages and written texts, for instance.

H. Herrero et al. (Eds.): RSME 2022, RSME 13, pp. 26–44, 2024.
https://doi.org/10.1007/978-3-031-62025-6_2

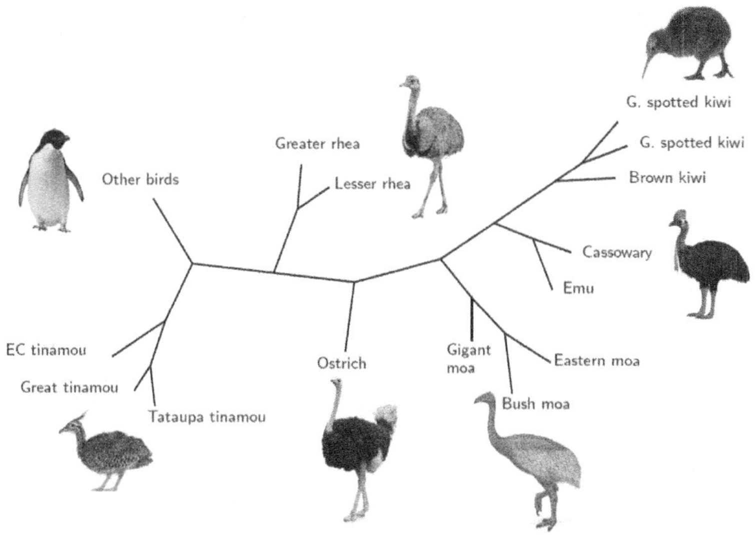

Fig. 1. A phylogenetic tree on several species of birds. Rheas, Kiwi, Ostrich, Cassowary and Emus, (extinct) Moas and Tinamous are paleognathus birds; they are known as ratites and among them only Tinamous can fly. It is still controversial whether this would be the correct phylogeney of these species, or alternatively, Tinamous should be placed within the clade of other Ratites, see [1,2]. Figure courtesy of Marina Garrote-López.

The main goal in phlyogenetic reconstruction is the following:

? Problem: phylogenetic reconstruction

Provide consistent methods that, given a collection of DNA sequences, produce the most plausible phylogenetic tree representing the evolution of the sequences.

Nowadays phylogenetics faces many different challenges. As there are more and more data available, there is a need of using more complex models fitting these data. On the contrary, phylogenetic reconstruction tools usually assume oversimplified models to make computations feasible. Moreover, the number of phylogenetic trees grows more than exponentially in the number of leaves (or species under study), so it is not possible to explore exhaustively the space of all phylogenetic trees. Thus, there is a need for new methods working for more complex models.

Here, the word "plausible" has a vague meaning on purpose, as it depends on the measure one wants to use. In order to reconstruct the phylogenetic tree, one usually models the evolution via a Markov process on the tree. This Markov process governs the substitution of nucleotides along the tree and is the basis of widely used methods such as maximum likelihood or Bayesian tools. At the beginning of this century, the emergence of the new discipline *algebraic statistics* (coined as in the book [5]) made possible the use of algebraic tools in statistical inference. These tools have been used in computational biology since the papers of L. Pachter and B. Sturmfels at Proceedings of the National Academy of Sciences, [6] and [7], and appear in many different areas nowadays.

In phylogenetics, the use of algebraic tools was initiated by E. Allman and J. Rhodes, and at present there is a solid community working in *algebraic phylogenetics*. The key observation is that Markov processes on trees are algebraic statistical models (i.e. parametric models for which the distribution is expressed as a polynomial in the parameters). From this, the study of algebraic varieties they define appears naturally. This has lead to relevant consequences from algebraic geometry in phylogenetics (see [8,9]) but also problems from phylogenetics have lead to important results in mathematics (see [10–13]) and other areas, see [14]. Recently, a refinement has been introduced in order to use semi-algebraic varieties instead of algebraic varieties, [15–17].

In this article we review the most important algebraic tools that are used in phylogenetic reconstruction. In Sect. 2 we give the main definitions, describe Markov processes on phylogenetic trees and how these lead to the natural study of the corresponding algebraic varieties. In Sect. 3 we provide the main equations that define these algebraic varieties. In Sect. 4 we state the semi-algebraic conditions that must be taken into account in phylogenetics. Finally, in Sect. 5 we explain how to use these tools in phylogenetic reconstruction. This is based on previous joint work with J. Fernández-Sánchez and M. Garrote-López in [17] and [16].

2 Markov Processes of Nucleotide Substitution

2.1 Phylogenetic Trees

We start with the basic definitions of phylogenetic trees, which can be found in the book by Mike Steel [18, chapter 1] for example.

A *tree* T is a connected acyclic graph with a collection of vertices $V(T)$ and edges $E(T)$. The *degree* of a vertex $v \in V(T)$ is the number of edges that are incident with v. The set of vertices of degree 1 are called *leaves* and those non-leaf vertices are called *interior* vertices.

Definition 1. *Let S be a finite set (in our setting S is usually a set of biological entities such as living species). An* (unrooted) phylogenetic tree on S *is a tree T with leaf set S whose interior vertices have degree at least three.*

In other words, a phylogenetic tree on S is a tree with no nodes of degree two together with a bijection between the set of leaves of T and the set S. The set S represents current species while the interior nodes represent their common ancestors, see Fig. 1 for an example. The set S will be usually understood from the context, so it will be often omitted.

Two phylogenetic trees T and T' on S are *isomorphic* if there is a graph isomorphism from T to T' that is the identity on the leaf set S. The topology of a phylogenetic tree refers to the isomorphism class of T as phylogenetic tree. See Fig. 2 for three classes of isomorphism of phylogenetic trees on $S = \{1, 2, 3, 4\}$, that will be represented as $T_{12|34}$, $T_{13|24}$, $T_{14|23}$.

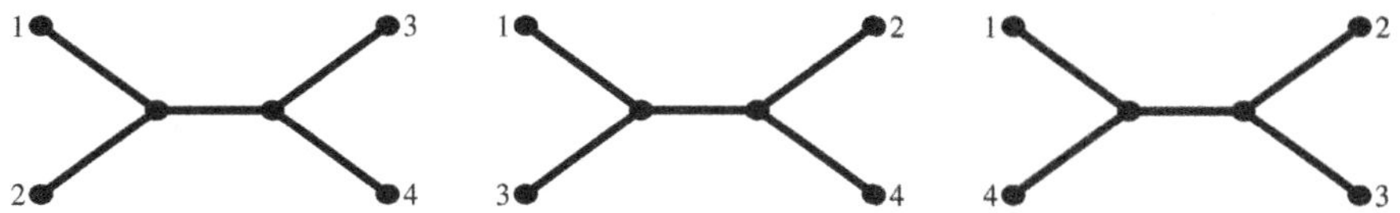

Fig. 2. Three non-isomorphic trees on the set of leaves 1, 2, 3, 4. These trees are denoted as $T_{12|34}$, $T_{13|24}$, and $T_{14|23}$ (from left to right) and, together with the star tree on four leaves, they represent all possible tree topologies on 1, 2, 3, 4.

We can *root* a phylogenetic tree by specifying a vertex r in $V(T)$ and directing all edges out of it. For reasons related to identifiability that will be clarified in the next subsection, in our setting we do not allow trees with degree two nodes.

There is another type of information that can be added on a phylogenetic tree: if there are weights assigned to the edges (or *branch lengths*), these usually represent an evolutionary distance between both ends of the edge. In this paper we do not take into account this kind of information.

2.2 Models of Nucleotide Substitution

In order to study the evolutionary relationships among DNA sequences, one specifies a mathematical model of substitution of nucleotides along a phylogenetic tree. Let T be an unrooted phylogenetic tree with leaves $S = \{1, \ldots, n\}$ and set an internal vertex r to play the role of the root so that we can direct the edges out of it (see for example, Fig. 3). The vertex r had a DNA sequence associated to it (a word on the alphabet A, C, G, T) and this sequence has randomly mutated to the DNA sequences that we observe nowadays at the leaves of the tree. To simplify things, it is common to assume that the nucleotides at different positions evolve independently of each other and following the same process (that is, sites at the DNA sequence are independent and identically distributed). Thus one only needs to model the evolution of a single nucleotide.

Assign a random variable X_i at each node $i \in V(T)$ and assume that the substitution of nucleotides along the tree follows a Markov process: the random variable at each node is conditionally independent of its non-descendant random variables given the random variable at its immediate parent node. Another way to describe this Markov process is via a parametric statistical model: let $\pi = (\pi_{\mathtt{A}}, \pi_{\mathtt{C}}, \pi_{\mathtt{G}}, \pi_{\mathtt{T}})$ be the distribution of nucleotides at the root of a tree ($\pi_{\mathtt{X}}$ is the probability of $\mathtt{X}$ at the DNA sequence at r) and let M^e be the matrix of conditional probabilities of substitution along edge $e : u \longrightarrow v$, that is,

$$M^e = \begin{array}{c} \\ \mathtt{A} \\ \mathtt{C} \\ \mathtt{G} \\ \mathtt{T} \end{array} \begin{array}{c} \begin{array}{cccc} \mathtt{A} & \mathtt{C} & \mathtt{G} & \mathtt{T} \end{array} \\ \begin{pmatrix} P(\mathtt{A}|\mathtt{A},e) & P(\mathtt{C}|\mathtt{A},e) & P(\mathtt{G}|\mathtt{A},e) & P(\mathtt{T}|\mathtt{A},e) \\ P(\mathtt{A}|\mathtt{C},e) & P(\mathtt{C}|\mathtt{C},e) & P(\mathtt{G}|\mathtt{C},e) & P(\mathtt{T}|\mathtt{C},e) \\ P(\mathtt{A}|\mathtt{G},e) & P(\mathtt{C}|\mathtt{G},e) & P(\mathtt{G}|\mathtt{G},e) & P(\mathtt{T}|\mathtt{G},e) \\ P(\mathtt{A}|\mathtt{T},e) & P(\mathtt{C}|\mathtt{T},e) & P(\mathtt{G}|\mathtt{T},e) & P(\mathtt{T}|\mathtt{T},e) \end{pmatrix} \end{array}.$$

Here $P(\mathtt{X}|\mathtt{Y}, e)$ denotes the conditional probability that nucleotide $\mathtt{Y}$ at the parent node u of e is substituted by nucleotide $\mathtt{X}$ at the child node v. Note that this matrix has non-negative entries and sum of rows equal to one; this is called a Markov matrix, transition matrix, or row stochastic matrix. With these parameters, the Markov process on the tree is specified by

$$Prob\left(\{X_w = \mathtt{X}_w\}_{w \in V(T)}\right) = \pi_{\mathtt{X}_r} \prod_{e:u \to v} M^e_{\mathtt{X}_u, \mathtt{X}_v}.$$

On a phylogenetic tree we do not have observations of the ancestral DNA sequences, so the random variables at the interior nodes of the tree are hidden. Thus, we obtain the probability $p_{\mathtt{X}_1 \ldots \mathtt{X}_n}$ of observing nucleotides $\mathtt{X}_i$ at leaf i by marginalizing the previous expression over the interior nodes:

$$p_{\mathtt{X}_1 \ldots \mathtt{X}_n} = \sum_{\substack{\mathtt{X}_u \in \{\mathtt{A},\mathtt{C},\mathtt{G},\mathtt{T}\} \\ u \in Int(T)}} \pi_{\mathtt{X}_r} \prod_{e:u \to v} M^e_{\mathtt{X}_u, \mathtt{X}_v}, \tag{1}$$

where $Int(T)$ denotes the set of interior nodes of T.

For example, for the phylogenetic tree of Fig. 3, we have

$$p_{\mathtt{ACCG}} = \sum_{\mathtt{X}_r} \sum_{\mathtt{X}_u} \pi_{\mathtt{X}_r} M^1_{\mathtt{X}_r,\mathtt{A}} M^2_{\mathtt{X}_r,\mathtt{C}} M^5_{\mathtt{X}_r,\mathtt{X}_u} M^3_{\mathtt{X}_u,\mathtt{C}} M^4_{\mathtt{X}_u,\mathtt{G}}.$$

The entries of π and M^e are parameters of this statistical model and, as we have just seen, the joint distribution of nucleotides at the leaves of the tree can be expressed as a polynomial function of these parameters. Therefore we have an algebraic statistical model and the following polynomial map sends each set of free parameters to the joint distribution $p^T = (p_{\mathtt{X}_1 \ldots \mathtt{X}_n})_{\mathtt{X}_1, \ldots, \mathtt{X}_n}$ of nucleotides at the leaves:

$$\begin{aligned} \varphi_T : \text{Free Parameters} &\longrightarrow \mathbb{R}^{4^n} \\ \left(\text{entries of } \pi, (M^e)_{e \in E(T)}\right) &\mapsto p^T = (p_{\mathtt{AA}\ldots\mathtt{A}}, p_{\mathtt{AA}\ldots\mathtt{C}}, p_{\mathtt{AA}\ldots\mathtt{G}}, \ldots, p_{\mathtt{TT}\ldots\mathtt{T}}). \end{aligned} \tag{2}$$

Any distribution p arising from a Markov process on the tree T is in the image of this map. Characterizing which distributions are in the image of this map is crucial for deciding whether a given distribution from real data is likely to have arisen as a Markov process on T or not. We come back to this problem in the next subsection, but for the moment we want to mention other evolutionary models.

The model presented above is commonly known as the *general Markov* (GM briefly) model, or the Barry-Hartigan model [19]. Depending on the biological data we are dealing with, simpler models can be considered. In what follows, we describe these simpler models as instances of *equivariant* models (see [20]).

Let G be a subgroup of the symmetric group $\mathfrak{S}_4$ on the set $\{\mathtt{A}, \mathtt{C}, \mathtt{G}, \mathtt{T}\}$, i.e. G is a group of permutations of these four elements. If in the parametric statistical model presented above we impose

(1) the distribution π is invariant by the action of G, that is, $(\pi_{\mathtt{A}}, \pi_{\mathtt{C}}, \pi_{\mathtt{G}}, \pi_{\mathtt{T}}) = (\pi_{g\mathtt{A}}, \pi_{g\mathtt{C}}, \pi_{g\mathtt{G}}, \pi_{g\mathtt{T}})$ for any $g \in G$, and
(2) each transition matrix M^e is a G-equivariant map: $M^e_{\mathtt{X},\mathtt{Y}} = M^e_{g\mathtt{X},g\mathtt{Y}}$ for any $g \in G$,

then we have a *G-equivariant* model of nucleotide substitution. Well known examples of G-equivariant models are (listed in decreasing order of complexity):

- The GM model above: it is a G-equivariant model if we take $G = \{id\}$. In this model we have 12 free parameters per edge (for each transition matrix) plus three free parameters for the distribution π at the root. The GM model evolving on the tree of Fig. 3 has 63 free parameters.
- Strand-symmetric model, see [21]: if $G = \langle(\mathtt{AT})(\mathtt{CG})\rangle$, the corresponding equivariant model preserves the symmetry between both strands of DNA molecules. In this case, as the distribution π at the root must be G-invariant, it satisfies $\pi_{\mathtt{A}} = \pi_{\mathtt{T}}$ and $\pi_{\mathtt{C}} = \pi_{\mathtt{G}}$.
- Kimura 3-parameter model, see [22]: when $G = \langle(\mathtt{AC})(\mathtt{GT}), (\mathtt{AG})(\mathtt{CT})\rangle$ the distribution at the root must be uniform $\pi = (\frac{1}{4}, \frac{1}{4}, \frac{1}{4}, \frac{1}{4})$ and the transition matrices satisfy symmetries that reflect the chemical properties of both groups of nucleotides, purines (adenine A and guanine G) and pyrimidines (cytosine C and thymine T). For this model and its submodels below, the uniform distribution π is the stationary distribution of all transition matrices of the model.
- Kimura 2-parameters, abbreviated as K80 (see [23]): $G = \langle(\mathtt{ACGT}), (\mathtt{AG})\rangle$; this is a submodel of K81 that considers all substitutions between purines and pyrimidines to be equally probably at each edge.

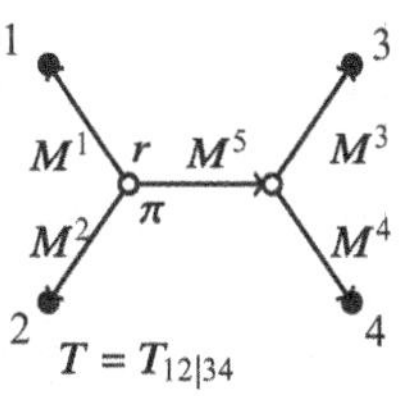

Fig. 3. A Markov process on the tree $T_{12|34}$.

- Jukes-Cantor model, abbreviated as JC69 (see [24]): this is the most simple model and corresponds to $G = \mathfrak{S}_4$. In this case the distribution at the root is uniform and the transition matrices at the edges are of the following type:

$$M^e = \begin{pmatrix} a_e & b_e & b_e & b_e \\ b_e & a_e & b_e & b_e \\ b_e & b_e & a_e & b_e \\ b_e & b_e & b_e & a_e \end{pmatrix},$$

where $a_e + 3b_e = 1$. For this model there is only one free parameter per edge (and there are no free parameters for the distribution at the root), thus this model on the tree of Fig. 3 has only five free parameters.

One of the first issues that one has to take into account when proposing a parametric model is *identifiability*, that is, whether all parameters of the model can be identified from data that perfectly fits the model. In our setting this translates to the following questions:

? Questions on identifiability

Let p be a distribution that has arisen as a Markov process on a phylogenetic tree T with parameters π and M^e.

- Can the tree T be identified solely from p?
- If the answer to the previous question is yes, can the parameters π and M^e be uniquely identified from p?

It is well known (see [18, §7.2.1]) that the answer to the first question is affirmative if p has arisen from *non-singular* parameters (that is, π has no zero entries and all matrices M^e are invertible and different from permutation matrices), which is a generic condition. Obviously, T can be recovered only up to isomorphism (so it is actually the tree topology that can be recovered from p); however, the vertex chosen to root the tree is not identifiable.

Furthermore, under the same conditions that guarantee an affirmative answer to the first question, there is an affirmative answer for the second: if we fix an interior vertex r to direct the tree, π and all transition matrices M^e can be recovered from p up to label swapping of states A, C, G, T at the interior nodes (see [25] for the GM model and [11] for the other G-equivariant models).

Thus the Markov models presented above are well-posed, in the sense that there is no overparametrization. Of course there are some sets of parameters that induce distributions at the leaves that can arise on differen trees, but this only occurs for particular parameters as we have seen (for example, in the tree of Fig. 3 we can set $M^5 = id$ and then the corresponding joint distribution at the leaves can also be obtained as a Markov process on any of the other two trees of Fig. 2).

If we had allowed nodes of degree two, then the answer to the identifiability questions would have been negative: if there is a node v of degree two (as the root in a binary rooted tree), one can only recover the product of the transition matrices of both edges incident with v (not the matrices separately).

There are more complex models that could be considered (for example allowing different sites to evolve according to different processes or allowing insertions and deletions of nucleotides and not only substitutions), but this is out of the scope of this survey.

2.3 Algebraic Geometry in Phylogenetics

If Δ is the standard simplex in $\mathbb{R}^{4^n}$, characterizing which distributions $p \in \Delta$ arise as a Markov process on a phylogenetic tree T can be useful in recovering the tree T from a given distribution. This is the main idea that leads to the use of algebraic geometry for phylogenetic reconstruction. The reader can have a look at the chapter [18, §8.3] for a good introduction to "phylogenetic algebraic geometry".

The main observation that leads to the use of an algebraic geometry perspective is the following. The image of the map φ_T defined in (2) "almost" fills an *algebraic variety* if we extend it to the complex field. An algebraic variety is the set of points where a collection of polynomials vanishes. The image of a polynomial map $\mathbb{C}^d \longrightarrow \mathbb{C}^{4^n}$ does not need to be an algebraic variety but it is always a *constructible set* (that is, in our case one needs to add to $Im\varphi_T$ some algebraic varieties of smaller dimension to obtain an algebraic variety). Let us call V_T the smallest algebraic variety containing the image of φ_T (i.e. the Zariski closure of $Im\varphi_T$). Then the polynomial equations that vanish on $Im\varphi_T$ are precisely those vanishing on V_T. Therefore, characterizing which distributions lie on the set $Im\varphi_T$ is equivalent to characterizing the intersection $V_T \cap \Delta$.

Summing up, the equations that define V_T almost determine the set $Im\varphi_T$. Finding these equations is not an easy task. Actually, there are infinitely many equations that vanish on V_T because the collection of equations that vanish on V_T form an *ideal* I_T in the ring of polynomials $\mathbb{R}[p_{\mathtt{AA...A}}, p_{\mathtt{AA...C}}, \ldots, p_{\mathtt{TT...T}}]$. According to Hilbert's basis theorem, this ideal is finitely generated. Biologists Cavender, Felsenstein and Lake in [26] and [27] were the first to introduce the idea of using polynomials that vanish on any distribution on a tree T. They called these polynomials in the ideal of V_T *phylogenetic invariants.* After a lot of efforts from algebraic geometers, collections of equations that define V_T have been found for trees on any number of leaves evolving under the evolutionary models introduced in the previous section, see [20,21,28,29]. Those wanting to play with these phylogenetic invariants can have a look at webpage

https://www.coloradocollege.edu/aapps/ldg/small-trees/small-trees_0.html which contains lists of generators for small trees, see [30].

The positive answer to the Questions on identifiability of the previous section ensure that any two of these varieties intersect properly (first question) and that the fibers of φ_T are zero-dimensional (second question). Therefore, the dimension of V_T (as algebraic variety, or as a manifold at the non-singular points) is equal to the dimension of the parameter space. For the GM model, this dimension equals $d = 3+12|E(T)|$, so the codimension of V_T is $4^n-3-12|E(T)|$, which is exponential on n (the number of edges of an n-leaved tree is bounded above by $2n-3$). The number of polynomials in a minimal system of generators of the ideal of V_T is too large to be used in practice: we need at least as many elements as the codimension to define a variety and we have just seen that this number is exponential on n. Moreover, it might be difficult to give a phylogenetic interpretation to some of these polynomials and to use them for phylogenetic reconstruction purposes. In Sect. 5 we explain other ways to use these ideals for phylogenetic reconstruction.

In what follows we give examples of phylogenetic invariants. For all the models and all trees, the following is a trivial phylogenetic invariant:

$$h : p_{\mathtt{AA\ldots A}} + p_{\mathtt{AA\ldots C}} + \cdots + p_{\mathtt{TT\ldots T}} - 1.$$

This polynomial vanishes on any point of $Im\varphi_T$ (and hence on the whole V_T) because probabilities must sum to one.

There are other phylogenetic invariants that depend only on the evolutionary model chosen but not on the tree structure. These are called *model invariants*. For example, for any tree on n-leaves evolving under the Jukes-Cantor model, the polynomial $p_{\mathtt{AA\ldots A}} - p_{\mathtt{CC\ldots C}}$ vanishes on any point $p^T \in Im\varphi_T$ (this can easily be seen using the symmetries of the transition matrices in JC69 model, and we can obtain other model invariants in the same way). Actually, if p is a distribution that has arisen from a G-equivariant model evolving on a tree T, then

$$p_{g\mathtt{X}_1,\ldots,g\mathtt{X}_n} = p_{\mathtt{X}_1,\ldots,\mathtt{X}_n}$$

for any $g \in G$ (i.e. p is G-invariant). Hence, $p_{g\mathtt{X}_1,\ldots,g\mathtt{X}_n} - p_{\mathtt{X}_1,\ldots,\mathtt{X}_n}$ are model invariants for any $g \in G$ and $\mathtt{X}_1,\ldots,\mathtt{X}_n \in \{\mathtt{A},\mathtt{C},\mathtt{G},\mathtt{T}\}^n$. These model invariants have been used in model selection in [31]: they have been implemented in a method that selects the G-equivariant model that best fits the data according to a statistical criterion.

The phylogenetic invariants that might be of interest in phylogenetic reconstruction are those that lie in I_T but not in $I_{T'}$ for some other tree T'. These are called *topology invariants*.

For example, for the JC69 or K80 model on the tree $T_{12|34}$, Lake [27] found the following linear phylogenetic invariants:

$$\begin{aligned} H_1 : &\quad p_{xyxy} + p_{xyzw} - p_{xyzy} - p_{xyxw} \\ H_2 : &\quad p_{xyyx} + p_{xywz} - p_{xyyz} - p_{xywx} \end{aligned} \qquad (3)$$

for any x, y, z, w in $\{\mathtt{A},\mathtt{C},\mathtt{G},\mathtt{T}\}$. It is not difficult to see that H_1 is not a phylogenetic invariant for $T_{13|24}$ and H_2 is not an invariant for $T_{14|23}$, so these polynomials are actually topology invariants for $T = T_{12|34}$. Lake used these two invariants to propose a method of phylogenetic reconstruction for quartet trees, without much success. It is not difficult to see why this method was not very successful: the variety V_T is not a linear variety so using only linear invariants may not give the best results in phylogenetic reconstruction. In the next section we explain how to obtain other topology invariants of larger degrees.

3 Invariants from Flattenings

For the moment we consider trees on four leaves $\{1, 2, 3, 4\}$. There are three possible (non-trivial) bipartitions of the set of leaves: 12|34, 13|24, 14|23. Let $p = (p_{AA...A}, p_{AA...C}, p_{AA...G}, \ldots, p_{TT...T})$ be a point in $\mathbb{R}^{4^4}$. Then we define the flattening of p according to the bipartition 12|34 as the matrix

$$flat_{12|34}(p) = \begin{matrix} \textit{states} \\ \textit{leaves} \\ 1,2 \end{matrix} \overset{\textit{states at leaves 3 and 4}}{\begin{pmatrix} p_{AAAA} & p_{AAAC} & p_{AAAG} & \cdots & p_{AATT} \\ p_{ACAA} & p_{ACAC} & p_{ACAG} & \cdots & p_{ACTT} \\ p_{AGAA} & p_{AGAC} & p_{AGAG} & \cdots & p_{AGTT} \\ \vdots & \vdots & \vdots & \vdots & \vdots \\ p_{TTAA} & p_{TTAC} & p_{TTAG} & \cdots & p_{TTTT} \end{pmatrix}}.$$

Note that this is a 16×16 matrix with rows labelled by the states AA, AC, ..., TG, TT at leaves 1 and 2 and columns labelled by the states at leaves 3 and 4. We can define $flat_{13|24}(p)$ and $flat_{14|23}(p)$ analogously. Another way of interpreting these matrices is by using tensors: p belongs to $\mathbb{R}^{4^4} \cong \mathbb{R}^4 \otimes \mathbb{R}^4 \otimes \mathbb{R}^4 \otimes \mathbb{R}^4$ and $flat_{12|34}(p)$ is the image of p via the isomorphism from $(\mathbb{R}^4 \otimes \mathbb{R}^4) \otimes (\mathbb{R}^4 \otimes \mathbb{R}^4)$ to the set of linear maps $L((\mathbb{R}^4 \otimes \mathbb{R}^4)^*, \mathbb{R}^4 \otimes \mathbb{R}^4)$.

Using expression (1) one can see that, if $T = T_{12|34}$ and $p = \varphi_T(\pi; M^1, M^2, M^3, M^4, M^5)$ belongs to $Im\varphi_T$, then

$$flat_{12|34}(p) = (M^1 \otimes M^2)^t flat_{12|34}(q) M^3 \otimes M^4$$

where $M \otimes N$ denotes the Kronecker product of matrices and $q = \varphi_T(\pi; Id, Id, Id, Id, M^5)$. It can be easily seen that $flat_{12|34}(q)$ is a 16×16 matrix whose unique non-zero entries are labelled by (XX, YY). Hence, $flat_{12|34}(q)$ has rank ≤ 4 and the same holds for $flat_{12|34}(p)$. This is the basis for the following result:

Theorem 1. *(Allman Rhodes, [32]) Let $p = \varphi_T(\pi; M^1, M^2, M^3, M^4, M^5)$ be a distribution arising from the GM model on $T = 12|34$. Then $rank(flat_{12|34}\, p) \leq 4$.*

Moreover, $rank(flat_{13|24}\, p)$ and $rank(flat_{14|23}\, p)$ are equal to 16 if M^i are invertible and π is strictly positive.

In other words, the 5×5 minors of $flat_{12|34}(p)$ are topology invariants for $T = T_{12|34}$. This result can be extended to larger trees by considering bipartitions $A|B$ of the set of leaves and flattening according to these bipartitions: the result would still refer to rank four matrices, as four is the number of states of the random variables we are considering at the nodes of the tree. Invariants from flattenings characterize whether a bipartition $A|B$ is present in the structure of the tree T giving rise to the distribution p: if T has an interior edge that partitions the set of leaves in $A|B$, then $\text{rank}(flat_{A|B}(p))$ is less than or equal to four.

An analogous version of this result has been proven for G-equivariant models in [33] by using techniques from representation theory. As a consequence, one can see that Lake's invariants in (3) appear from these rank conditions under the JC69 model. Moreover, in the quoted paper it is proven that these rank conditions are enough to define a variety V_{T_0} inside the union of varieties $\cup_{T\in\mathcal{T}} V_T$ (where $\mathcal{T}$ is the set of trees on the set of leaves $\{1,\ldots,n\}$):

Theorem 2. *(Casanellas, Fernández-Sánchez,[33]) Consider any G-equivariant model. For each tree topology $T \in \mathcal{T}$, there exists a dense open subset $U_T \subseteq V(T)$ such that if a point p belongs to $\bigcup_{T\in\mathcal{T}} U_T$, then p belongs to V_{T_0} if and only if the rank of all $flat_{A|B}(p)$ is less than or equal to four for all bipartitions induced from interior edges of T_0.*

Roughly speaking, if one assumes that p has arisen as a Markov process on some phylogenetic tree (with generic parameters), considering invariants from flattenings is enough to detect the tree topology. As assuming that p has arisen on some phylogenetic tree with generic parameters (i.e. $p \in \bigcup_{T\in\mathcal{T}} U_T$) is what all common phylogenetic reconstruction methods do, this theorem says that considering rank conditions from flattenings is sufficient for tree topology reconstruction via algebraic geometry.

> **Recall**

Assuming that data comes from a phylogenetic tree, it is enough to consider rank conditions from flattenings to recover the tree (see Theorem 2).

In Sect. 5 below, we shall see an alternative way to use these rank conditions in practice.

4 Semi-algebraic Constraints

As the reader may have realized, when passing from the Markov process on a tree T to the map φ_T in (2), we are eluding the fact that the parameters are probabilities. Let us call φ_T^+ the restriction of φ_T to *stochastic parameters* (non-negative parameters that sum to one) and let $V_T^+ = Im\varphi_T^+$. The map φ_T^+ defines a *semi-algebraic variety*, that is, it can be described by polynomial equations and polynomial inequalities. In [34] we examined whether it is relevant to take this into account and we concluded that there are biologically realistic situations where considering the whole algebraic variety (and not only this part V_T^+) can be misleading. These situations are actually the typical scenarios where most phylogenetic reconstruction methods fail.

In [15], Allman, Rhodes and Taylor have characterized those distributions that lie in V_T^+. To this end, given a tensor $p \in \mathbb{R}^4 \otimes \mathbb{R}^4 \otimes \mathbb{R}^4 \otimes \mathbb{R}^4$ they consider some transformations $\tilde{p}$ of p obtained by multiplying trivial flattenings of p (considering bipartitions of one leaf against the others) by matrices coming from the marginalization of p over two indices. If $p = \varphi_T(\pi; M_1, M_2, M_3, M_4, M_5)$ is in

$Im\varphi_T$, $T = T_{12|34}$, these transformations produce a point $\tilde{p}$ also in $Im\varphi_T$ but with different transition matrices: $\tilde{p} = \varphi_T(\pi; M, M, N, N, M_5)$ where $M = M_1$ or M_2 and $N = M_3$ or M_4. We state here their result in the case of trees with four leaves.

Theorem 3. *(Allman, Rhodes and Taylor, 2014) A point $p \in \mathbb{R}^{4^4}$ arises from a Markov process on the phylogenetic tree $T_{12|34}$ with stochastic parameters if and only if:*

(i) $flat_{12|34}(p)$ has rank ≤ 4,
(ii) the marginalizations $p_{+\ldots}$ and $p_{\ldots+}$ arise from stochastic parameters, and
(iii) $flat_{13|24}(\tilde{p})$ is positive semi-definite, where $\tilde{p}$ is any of the transformations mentioned above.

In the next section we will explain how to use this result in phylogenetic reconstruction.

5 Phylogenetic Reconstruction

We come back to the problem of phylogenetic reconstruction. Our input data are DNA sequences of length N given in the form of an *alignment.* An alignment of n DNA sequences is an $n \times N$ array whose rows correspond to the DNA sequences and whose columns represent nucleotides that have evolved from the same nucleotide at the common ancestor of the sequences. For example, in Fig. 4 we have an alignment of four DNA sequences; the first nucleotide in each sequence corresponds to a certain nucleotide at the common ancestor of these sequences (probably an A as well).

s_1 : AACTTCGAGGCTTACC
s_2 : AAGGTCGATGCTCACC
s_3 : AACGTCTATGCTCACC
s_4 : GACGCCGATGCTCATC

Fig. 4. An alignment of four DNA sequences

As we assumed that all positions in a DNA sequence evolve independently and in the same way, an alignment can be thought of as N independent samples from a multinomial distribution $p = (p_{\mathtt{AA\ldots A}}, p_{\mathtt{AA\ldots C}}, \ldots, p_{\mathtt{TT\ldots T}})$. This distribution can be estimated from the relative frequencies $f = (f_{\mathtt{AA\ldots A}}, \ldots, f_{\mathtt{TT\ldots T}})$ of each column $\mathtt{X}_1, \ldots \mathtt{X}_n$ in the alignment, and $f \sim p$ when N tends to infinity. If these sequences had evolved according to a Markov process on a tree T (for certain parameters), then p would belong to $Im\varphi_T^+$ and f would be close to V_T^+.

The aim is to use the results in the previous sections that characterized distributions in V_T and V_T^+ to design methods that reconstruct T from the vector of relative frequencies f. A phylogenetic reconstruction method is called *statistically consistent* if it outputs T with probability one when the alignment has been generated by $p \in Im\varphi_T^+$ and we let the length of the alignment tend to infinite. This is the least we should require to a phylogenetic reconstruction method. Designing a reconstruction method that is statistically consistent is not so difficult, the difficult part is to make it reach convergence fast enough because sequences at our disposal are often not very long.

One of the most common phylogenetic reconstruction methods is *maximum likelihood estimation.* Given an alignment D and an evolutionary model, the method seeks to obtain the tree topology T_0 and the substitution parameters $\hat{\theta} = (\pi; \{M^e\}_e)$ which maximize $Prob(D|T, \theta)$ among all possible phylogenetic trees T and substitution parameters θ. To this end, first the maximum likelihood estimate of the substitution parameters is obtained separately for each tree topology T (using some of the available optimization methods) and then one chooses the tree topology and the parameter estimates which maximize the likelihood among all tree topologies.

This method has a clear drawback: the number of (isomorphism classes of) trivalent phylogenetic trees on n leaves is $(2n-5)!!$, which grows exponentially in n, so that it becomes unfeasible to do an exhaustive search through all tree topologies for more than 20 leaves (even with nowadays computational capacity). The vast majority of phylogenetic reconstruction software use some branch and bound algorithm and only cover a small part of the tree space. On the other hand, numerical optimization methods do not guarantee a global maximum in general and, moreover, it is known that there are multiple local maximum for biologically relevant parameters.

By far, the most used phylogenetic reconstruction method is Neighbor-Joining [35]. This a *distance-based method*, that is, all the information from an alignment on a set of species $\{1, \ldots, n\}$ is condensed into a *dissimilarity function* $d : \{1, \ldots, n\} \times \{1, \ldots, n\} \longrightarrow \mathbb{R}_{\geq 0}$ (symmetric and with zero diagonal entries). This dissimilarity function is intended to approximate the evolutionary distance between pairs of species or, in other words, it should account for the amount of elapsed substitutions between both species. Obviously, not all substitutions that have occurred during evolution can be observed in the contemporary species sequences (for instance, there may be an A mutating to T and finally coming back to A in the nowadays species) and the dissimilarity function has to take this into account. For example, the *Jukes-Cantor distance* between two DNA sequences defined as $-\frac{3}{4}\ln(1-\frac{4}{3}s)$, where s is the fraction of nucleotides that differ in both sequences. It approximates the amount of substitutions (observed and unobserved) between the species if these have evolved under the Jukes-Cantor model.

Given a dissimilarity function d, the first step in the Neighbor-Joining algorithm chooses two species x and y minimizing function $D_{x,y} = d(x,y) - \frac{1}{n-2}\sum_z(d(x,z) + d(y,z))$, seeking to minimize their distance but maximize the average distance to the other species. These two species are joined on a cherry (that is, two leaves joined by two edges and an interior node) and the interior node is treated as a new species substituting the former x and y. In this way the number of species is decreased at each step and the function D is redefined accordingly.

This algorithm produces the correct phylogenetic tree if the input dissimilarities can be realized as sums of lengths on the path between leaves on a phylogentic tree. However, when dealing with biological sequences, their estimated distances do not correspond to the branch lengths of any particular tree, and the tree constructed by Neighbor-joining algorithm may not have a realistic biological interpretation. In spite of this, it is one of the most widely used method and, as there is no need to search through the whole space of phylogenetic trees, it is used to produce phylogenetic trees for large number of species.

In the next section we explain of to use an algebraic geometry approach to produce alternative methods of phylogenetic reconstruction.

5.1 Reconstruction of Quartets

We start by reviewing algebraic reconstruction methods for quartet trees and in the next subsection we explain how to get to larger trees.

One of the most simple ways to use rank conditions (Theorem 1) in phylogenetic reconstruction is via the singular value decomposition of the flattening matrix [36]. A *singular value decomposition* of an $m \times n$ matrix M is a decomposition $M = UDV^t$ where U, V are orthogonal matrices and D is an $m \times n$ matrix with off-diagonal entries equal to zero and ordered diagonal entries $D_{1,1} = \sigma_1 \geq \sigma_2 \geq \cdots \geq D_{r,r} = \sigma_r > 0$, $D_{i,i} = 0$ for $i > r = rankM$. The elements $\sigma_1, \ldots, \sigma_r$ are called singular values and are uniquely determined by M. The Eckart-Young theorem [37] states that the distance (in Frobenius norm) from a matrix M to the set of matrices of rank $\leq k$ is given by

$$\delta_k(M) = \sqrt{\sigma_{k+1}^2 + \cdots + \sigma_r^2}.$$

According to Theorem 1, if $T = 12|34$ and $p \in Im\varphi_T$ has been generated from generic parameters, we have $\delta_4(flat_{12|34}(p)) = 0$ and $\delta_4(flat_{13|24}(p)) > 0$, $\delta_4(flat_{14|23}(p)) > 0$. This idea was explored in [38] to propose a method of phylogenetic reconstruction: given the vector f of relative frequencies obtained from an alignment of DNA sequences, compute the triplet of scores

$$s_{12|34}(f) = \delta_4(flat_{12|34}(f)), \quad s_{13|24}(f) = \delta_4(flat_{13|24}(f)),$$

$$s_{14|23}(f) = \delta_4(flat_{14|23}(f))$$

and output the tree $T_{A|B}$ if $s_{A|B}(f)$ is the smallest.

In [39] this method was further exploited and a modification was proposed to make it more efficient and more robust. Despite of this improvement, methods based solely on rank conditions seem to need large alignments to outperform other reconstruction methods. That is, the method presented is statistically consistent but it converges slowly to the correct tree.

As argued in [34], methods based only on algebraic conditions may have problems in reconstructing the correct tree from small samples because they do not take into account that parameters generating the alignment must be positive. However, incorporating semi-algebraic conditions does not seem an easy task. Out of the three conditions stated in Theorem 3, we are mostly interested in the (i) and (iii), as the second one is not directly related to the tree structure. The first is the rank conditions we discussed above. In order to incorporate the third condition we borrow the following result from linear algebra:

Theorem 4. *(Higham [40]) Let M be an $n \times n$ matrix. Consider its closest symmetric matrix $S = \frac{M+M^t}{2}$ and the polar decomposition of S, $S = UH$, where U is orthogonal and H is a positive semidefinite matrix. Then $psd(M) = \frac{S+H}{2}$ is the closest positive semidefinite matrix to M (in Frobenius norm).*

This theorem may allow us to compute how far is $flat_{13|24}(\tilde{p})$ from the set of positive semidefinite matrices. However, the goal is to combine conditions (i) and (iii) in Theorem 3 into a single score. To this aim, we proved in [41] that the rank of $psd(M)$ is always smaller than or equal to the rank of M. With all this in mind, in [16] we came out with the following method: given the vector of relative frequencies f of an alignment, compute all transformations of the vector mentioned in Theorem 3 (see [16] for details) and for each transformation $\tilde{f}$ compute

$$s_{12|34}(\tilde{f}) := \frac{\min\left\{\delta_4\left(psd\left(flat_{13|24}\left(\tilde{f}\right)\right)\right), \delta_4\left(psd\left(flat_{14|23}\left(\tilde{f}\right)\right)\right)\right\}}{\delta_4\left(psd\left(flat_{12|34}\left(\tilde{f}\right)\right)\right)}. \quad (4)$$

Then we define the score $s_{12|34}(f)$ of the alignment as the average of these scores and define $s_{13|24}(f)$, $s_{13|24}(f)$ accordingly. Note that for a distribution $p \in Im\varphi_T^+$ with $T = T_{12|34}$, we have $s_{12|34}(p) = \infty$. Indeed, according to Theorem 3(iii), $flat_{13|24}(\tilde{p})$ is positive semidefinite, so

$$rank\left(psd\left(flat_{13|24}(\tilde{p})\right)\right) = rank\left(flat_{13|24}(\tilde{p})\right).$$

As any transformation $\tilde{p}$ of p belongs to V_T, this rank is greater than 4 (for generic parameters). This proves that the numerator is different than zero. On the other hand, as mentioned above, the rank of $psd\left(flat_{12|34}(\tilde{p})\right)$ is bounded above by the rank of $flat_{12|34}(\tilde{p})$, which is smaller than or equal to four (because $\tilde{p} \in V_T$). Therefore, the denominator of 4 is zero and $s_{12|34}(p) = \infty$.

With the same arguments we can prove that the other two scores $s_{13|24}(p)$ and $s_{14|23}(p)$ are zero. Thus, we can consider a method that outputs the tree $T_{A|B}$ for which $s_{A|B}(f)$ is largest. This is called the SAQ (for Semi-Algebraic Quartet) method in [16]. The previous argument proves that SAQ is a statistically consistent method. Moreover, it is much more efficient than previous algebraic methods and outperforms maximum-likelihood and neighbor-joining for alignments of length at least 500 generated under the GM model (whereas a method that only considers rank conditions needs at least 10000 sites to beat other methods).

5.2 From Quartets to Larger Trees

So far we have just seen algebraic methods that reconstruct only trees with four leaves (quartets). We are interested now in reconstructing phylogenetic trees with n leaves. Some methods, like Neighbor-Joining introduced above, build directly a tree with n leaves from an alignment of DNA sequences. But other methods built an n-leaved tree out of smaller pieces, namely quartet trees. These are called *quartet-based methods.*

We briefly describe one of these methods, Weight Optimization (WO for short), see [42]. WO needs as input a triplet of weights for each four species. That is, starting with an alignment of n DNA sequences, we need to consider all 4-tuples i, j, k, l of the sequences and use a reconstruction method to weight the three possible quartet trees: $w_{ijkl} = (w_{ij|kl}, w_{ik|jl}, w_{il|jk})$, each weight expressing the reliability of the corresponding quartet tree. For example, we can use SAQ to produce input weights for WO by defining:

$$w_{ijkl} = \left(\frac{s_{T_{ij|kl}}(f)}{s(f)}, \frac{s_{T_{ik|jl}}(f)}{s(f)}, \frac{s_{T_{il|jk}}(f)}{s(f)} \right)$$

where $s(f) = s_{T_{ij|kl}}(f) + s_{T_{ik|jl}}(f) + s_{T_{il|jk}}(f)$. In this way, weights are normalized between 0 and 1.

Then WO randomly chooses a starting 4-tuple i, j, k, l and dynamically defines the species addition order seeking to maximize the total sum of weights at each step: the added species at each step is selected and placed at the edge that gives the largest possible increase of weight. WO is known to reconstruct the correct tree if the quartets are correctly weighted. We can use SAQ to weight quartet trees but we could also use a method based only on rank conditions or on maximum likelihood strategies. As WO is highly dependant on the initial quartet, usually the method is run for several trials, say 100, so that 100 n-leaved trees are produced. These trees might not coincide and might be incompatible but a *consensus tree* can be built out of them (following a chosen criterion, such as the majority rule consensus tree, see [43]).

We ran WO with SAQ on the DNA sequences from Ratites used in [2] for 100 trials. The majority rule consensus tree obtained is the tree depicted in Fig. 1 (see [17]). Therefore, our methods support the hypothesis that Tinamous evolved separately from the rest of Ratites (see the discussion at the legend of the figure).

6 Discussion

In this report we have presented probabilistic models of nucleotide substitution that allow approaching phylogenetic reconstruction from an algebraic point of view. We have introduced some of the main techniques that are used in the development of algebraic and semi-algebraic tools in phylogenetics and we have seen how these tools can be used in the reconstruction of the tree topology of a set of DNA sequences.

Still, there are problems for which algebraic phylogenetics has not been fully developed yet. For instance, it would be interesting to provide tree reconstruction methods for G-equivariant models, not only for the general Markov model. This would be specially interesting in the case of protein data, as the general Markov model of amino acid substitution seems too general to treat these data: it will involve 380 parameters per edge whereas current models used by biologists only involve one. Developing algebraic tools for models in between these two would certainly be of interest.

Also, we have assumed that sites in the sequences were identically distributed. Considering mixtures of distributions is a way of relaxing these assumptions. Mixtures of distributions are just convex combinations of distributions, and therefore also allow an algebraic approach. Further studying mixtures for different evolutionary models can lead to more flexibility in the models where algebraic tools can be applied. We need to mention that the tools presented here work for gene trees, and one has to combine them with a coalescent model in order to infer species trees.

On the other hand, we have mentioned how to reconstruct trees from quartets inferred by algebraic tools. It would be desirable to extend algebraic techniques to develop methods of reconstruction of large trees directly, not passing through quartets.

Acknowledgments. The author would like to thank her coauthors Jesús Fernández-Sánchez and Marina Garrote-López for all the joint work described here. The author has been partially supported by PID2019-103849GB-I00 and CEX2020-001084-M of AEI, Government of Spain.

References

1. Phillips, M.J., Gibb, G.C., Crimp, E.A., Penny, D.: Tinamous and Moa flock together: mitochondrial genome sequence analysis reveals independent losses of flight among ratites. Syst. Biol. **59**(1), 90–107 (2009)
2. Vera-Ruiz, V.A., Robinson, J., Jermiin, L.S.: A likelihood-ratio test for lumpability of phylogenetic data: is the Markovian property of an evolutionary process retained in recoded DNA? Syst. Biol. **71**(3), 660–675 (2021)
3. Jermiin, L.S., Catullo, R.A., Holland, B.R.: A new phylogenetic protocol: dealing with model misspecification and confirmation bias in molecular phylogenetics. NAR Genom. Bioinform. **2**(2) (2020)
4. Makarenkov, V., Mazoure, B., Rabusseau, G., Legendre, P.: Horizontal gene transfer and recombination analysis of SARS-CoV-2 genes helps discover its close relatives and shed light on its origin. BMC Ecol. Evol. **21**, 2730–7182 (2021)
5. Pistone, G., Riccomagno, E., Wynn, H.P.: Algebraic Statistics: Computational Commutative Algebra in Statistics. Chapman and Hall/CRC, Boca Raton (2000)
6. Pachter, L., Sturmfels, B.: Parametric inference for biological sequence analysis. Proc. Natl. Acad. Sci. **101**(46), 16138–16143 (2004)
7. Pachter, L., Sturmfels, B.: Tropical geometry of statistical models. Proc. Natl. Acad. Sci. **101**(46), 16132–16137 (2004)

8. Allman, E.S., Ané, C., Rhodes, J.A.: Identifiability of a Markovian model of molecular evolution with gamma-distributed rates. Adv. Appl. Probab. **40**(1), 229–249 (2008)
9. Sumner, J.G., Jarvis, P.D., Fernández-Sánchez, J., Kaine, B.T., Woodhams, M.D., Holland, B.R.: Is the general time-reversible model bad for molecular phylogenetics? Syst. Biol. **61**(6), 1069–1074 (2012)
10. Casanellas, M., Fernández-Sánchez, J., Roca-Lacostena, J.: The embedding problem for Markov matrices. Publicacions Matemàtiques **67**(1), 411–445 (2023)
11. Casanellas, M., Fernández-Sánchez, J., Michałek, M.: Local equations for equivariant evolutionary models. Adv. Math. **315**, 285–323 (2017)
12. Friedland, S., Gross, E.: A proof of the set-theoretic version of the salmon conjecture. J. Algebra **356**(1), 374–379 (2012)
13. Michałek, M., Ventura, E.: Phylogenetic complexity of the kimura 3-parameter model. Adv. Math. **343**, 640–680 (2019)
14. Casanellas, M., Garrote-López, M., Zwiernik, P.: Identifiability in robust estimation of tree structured models. Bernoulli J. **30**(1), 1–21 (2024)
15. Allman, E.S., Rhodes, J.A., Taylor, A.: A semialgebraic description of the general Markov model on phylogenetic trees. SIAM J. Discret. Math. **28**(2), 736–755 (2014)
16. Casanellas, M., Fernández-Sánchez, J., Garrote-López, M.: SAQ: semi-algebraic quartet reconstruction method. IEEE/ACM Trans. Comput. Biol. Bioinf. **18**(6), 2855–2861 (2021)
17. Casanellas, M., Fernández-Sánchez, J., Garrote-López, M., Sabaté-Vidales, M..: Designing weights for quartet-based methods when data is heterogeneous across lineages. Bull. Math. Biol. **85**, 68 (2023)
18. Steel, M.A.: Phylogeny: Discrete and Random Processes in Evolution. SIAM-Society for Industrial and Applied Mathematics, Philadelphia, PA, USA (2016)
19. Barry, D., Hartigan, J.A.: Asynchronous distance between homologous DNA sequences. Biometrics **43**(2), 261 (1987)
20. Draisma, J., Kuttler, J.: On the ideals of equivariant tree models. Math. Ann. **344**(3), 619–644 (2009)
21. Casanellas, M., Sullivant, S.: The strand symmetric model. In: Pachter, L., Sturmfels, B. (eds.) Algebraic Statistics for Computational Biology, chap. 16, pp. 305–321. Cambridge University Press, Cambridge (2005)
22. Kimura, M.: Estimation of evolutionary distances between homologous nucleotide sequences. Proc. Natl. Acad. Sci. **78**(1), 454–458 (1981)
23. Kimura, M.: A simple method for estimating evolutionary rates of base substitutions through comparative studies of nucleotide sequences. J. Mol. Evol. **16**(2), 111–120 (1980)
24. Jukes, T.H., Cantor, C.R.: Evolution of protein molecules. Mamm. Protein Metab. **3**, 21–132 (1969)
25. Chang, J.T.: Full reconstruction of Markov models on evolutionary trees: identifiability and consistency. Math. Biosci. **137**(1), 51–73 (1996)
26. Cavender, J.A., Felsenstein, J.: Invariants of phylogenies in a simple case with discrete states. J. Classif. **4**(1), 57–71 (1987)
27. Lake, J.A.: A rate-independent technique for analysis of nucleic acid sequences: evolutionary parsimony. Mol. Biol. Evol. **4**, 167–191 (1987)
28. Allman, E.S., Rhodes, J.A.: Phylogenetic ideals and varieties for the general Markov model. Adv. Appl. Math. **40**(2), 127–148 (2008)
29. Sturmfels, B., Sullivant, S.: Toric ideals of phylogenetic invariants. J. Comput. Biol. **12**(2), 204–228 (2005)

30. Casanellas, M., Garcia, L.D., Sullivant, S.: Catalog of small trees. In: Pachter, L., Sturmfels, B. (eds.) Algebraic Statistics for Computational Biology, chap. 15, pp. 305–321. Cambridge University Press, Cambridge (2005)
31. Kedzierska, A.M., Drton, M., Guigó, R., Casanellas, M.: SPIn: model selection for phylogenetic mixtures via linear invariants. Mol. Biol. Evol. **29**(3), 929–937 (2012)
32. Allman, E.S., Rhodes, J.A.: Phylogenetic invariants for the general Markov model of sequence mutation. Math. Biosci. **186**(2), 113–144 (2003)
33. Casanellas, M., Fernández-Sánchez, J.: Relevant phylogenetic invariants of evolutionary models. Journal de Mathématiques Pures et Appliquées **96**(3), 207–229 (2011)
34. Casanellas, M., Fernández-Sánchez, J., Garrote-López, M.: Distance to the stochastic part of phylogenetic varieties. J. Symb. Comput. **104**, 653–682 (2021)
35. Saitou, N., Nei, M.: The neighbor-joining method: a new method for reconstructing phylogenetic trees. Mol. Biol. Evol. **4**, 406–425 (1987)
36. Meyer, C.D.: Matrix analysis and applied linear algebra. Society for Industrial and Applied Mathematics, Philadelphia, PA, USA (2000)
37. Eckart, C., Young, G.: The approximation of one matrix by another of lower rank. Psychometrika **1**(3), 211–218 (1936)
38. Eriksson, N.: Tree construction using singular value decomposition. In: Pachter, L., Sturmfels, B. (eds.) Algebraic Statistics for Computational Biology, chap. 19, pp. 347–358. Cambridge University Press, Cambridge (2005)
39. Fernández-Sánchez, J., Casanellas, M.: Invariant versus classical approach when evolution is heterogeneous across sites and lineages. Syst. Biol. **65**, 280–291 (2016)
40. Higham, N.J.: Matrix nearness problems and applications. In: Applications of Matrix Theory, vol. 22 (1989)
41. Casanellas, M., Fernández-Sánchez, J., Garrote-López, M.: The inertia of the symmetric approximation for low-rank matrices. Linear Multilinear Algebra **66**(11), 2349–2353 (2018)
42. Ranwez, V., Gascuel, O.: Quartet-based phylogenetic inference: improvements and limits. Mol. Biol. Evol. **18**(6), 1103–1116 (2001)
43. Barrett, M., Donoghue, M.J., Sober, E.: Against consensus. Syst. Zool. **40**(4), 486–493 (1991)

A Sample of Problems Where Ulrich Bundles Show up

Laura Costa(✉)

Facultat de Matemàtiques i Informàtica, Departament de Matemàtiques i Informàtica, Universitat de Barcelona, Gran via de les Corts Catalanes 585, 08007 Barcelona, Spain
costa@ub.edu

Abstract. The goal of these notes is to show that Ulrich bundles are key objects in order to achieve the solution of very different problems. They were born in the realm of commutative algebra and we will see that they play an important role for instance in the solution of geometrical problems, they have connections with mathematical physics and have been used to better understand a problem in computer vision.

1 Introduction

The theory of vector bundles on projective varieties is one of the mainstreams in Algebraic Geometry. They appear in the 60's with the works of Schwarzenberger but it was during the 1970's and 1980's, that they received a major boost. Among other things, vector bundles offer a valuable point of view in order to better understand the geometry and the topology of algebraic varieties.

In general, since the whole category of vector bundles on a projective variety is usually unwieldy, one has to impose some restrictions on the families of vector bundles under consideration. In this sense, basically two approaches have been proposed: the geometric one based on the notion of stability and a second one that deals with the cohomological properties of vector bundles. In these notes we are focus on this last approach. Mainly, we will focus the attention on Ulrich bundles. Ulrich bundles form a subfamily that recently has been the object of intense research. They can be roughly defined as those bundles without intermediate cohomology on a projective variety that have the largest permitted number of global sections.

The existence of an Ulrich bundle on a d-dimensional projective variety X has strong consequences for the underlying variety X. To mention just a couple, it implies that the cone of cohomology tables of vector bundles on X is the same as the one of the projective space $\mathbb{P}^d$. On the other hand, it also implies that its associated Cayley-Chow form has a particularly simple presentation. Despite all of these constraints, it has been asked by Eisenbud and Schreyer whether any projective variety supports an Ulrich bundle. Prompted by this question, the theory of Ulrich bundles has became a very active area of research within algebraic geometry with a constant supply of new achievements.

H. Herrero et al. (Eds.): RSME 2022, RSME 13, pp. 45–61, 2024.
https://doi.org/10.1007/978-3-031-62025-6_3

The goal of these notes is to provide examples that show how Ulrich bundles can appear in very different context and related with, a priori, very different questions. To this end, in Sect. 1 we will see that their origins are in the context of commutative algebra and related with Maximally Generated Maximal Cohen Macaulay Modules. In Sect. 2 we will see how Ulrich bundles are the key to solve a problem posted by geometers looking for determinantal representations of hypersurfaces. In Sect. 3 we will show a relation between the existence of certain Ulrich bundles and the existence of instanton bundles, an object deeply related with mathematical physics. We end these notes pointing out how Ulrich bundles has shown in computer vision.

These notes are the expanded version of the content of a talk prepared for the *Congreso Bienal de la RSME 2022*. I want to express my gratitude to the organizers for the opportunity to participate in this nice meeting. This was an expository talk about questions related with Ulrich bundles but in no case these notes pretend to be a summary of all of what is known about Ulrich bundles. Since they are a compilation of different aspects of Ulrich bundles done during the last decades, the material presented is not new. All details omitted here can be found in the book [19], the papers [18,27]. I want to knowledge the work of the authors of these references and many other that have contributed to the study of Ulrich bundles.

Notation: Throughout these notes $\mathbf{k}$ is an algebraically closed field of characteristic 0. Given a $(n+1)$-dimensional $\mathbf{k}$-vector space V we set $\mathbb{P}^n = \mathbb{P}(V)$ the projective space of hyperplanes of V. The associated coordinate ring of $\mathbb{P}^n$ is $R := \oplus_{i\geq 0} S^i V \cong \mathbf{k}[x_0, \ldots, x_n]$. If $X \subset \mathbb{P}^n$ is a closed subscheme of codimension c, we denote the ideal sheaf of X by $\mathcal{I}_X$ and the homogeneous saturated ideal by $I(X) = \oplus_{t\in\mathbb{Z}} \mathrm{H}(\mathcal{I}_X(t))$. R_X stands for the homogeneous coordinate ring of X defined as $\mathbf{k}[x_0, \cdots, x_n]/I(X)$.

For any coherent sheaf $\mathcal{E}$ on a projective variety $(X, \mathcal{O}_X(H))$ we are going to denote the twisted sheaf $\mathcal{E}\otimes\mathcal{O}_X(tH)$ by $\mathcal{E}(tH)$ or simply $\mathcal{E}(t)$. As usual, $\mathrm{H}^i(X,\mathcal{E})$ stands for the cohomology groups and $\mathrm{h}^i(X,\mathcal{E})$ for their dimension. If there is no confusion, we write $\mathrm{H}^i(\mathcal{E})$ instead of $\mathrm{H}^i(X,\mathcal{E})$ and by $\mathrm{h}^i(\mathcal{E})$ its dimension. In addition, we denote by $\mathrm{H}^i_* \mathcal{E} := \oplus_{t\in\mathbb{Z}} \mathrm{H}^i(\mathcal{E}(t))$.

2 Origins in Commutative Algebra

Given a local Cohen-Macaulay ring $(S, \mathfrak{m})$, it has been of high interest to give criteria for S being a Gorenstein ring. A possible approach would be to understand which kind of modules with some sort of extremal behaviour supports a local Gorenstein ring S. For instance, one could be interested in S-modules M with a large minimal number of generators. For M an S-module, it was proved that M is Cohen-Macaulay if and only if $\dim(M/\underline{x}M) = e(S)\,\mathrm{rank}(M)$ where $\underline{x}$ is a maximal regular sequence and $e(S)$ is the multiplicity of the ring. Therefore, it turns out that there is a natural bound on the cardinality of a minimal set of generators of a Cohen-Macaulay module M:

$$\mu(M) = \dim(M/\mathfrak{m}M) \leq \dim(M/\underline{x}M) = e(S)\,\mathrm{rank}(M).$$

The modules with large minimal number of generators turned out to have a great importance and Ulrich raised the following question (see [37, page 26]):

Question 2.1. Let S be a local Cohen-Macaulay ring with positive dimension and infinite residue class field. Does S support a Cohen-Macaulay module M with positive rank and maximal permitted cardinality of a minimal set of generators (namely, $\mu(M) = e(S)\,\mathrm{rank}(M)$)?

In [8], where this kind of modules where called Maximally Generated Maximal Cohen-Macaulay modules (MGMCM), their existence was proven on certain classes of rings. When we move on the realm of homogenous rings $S = \mathbf{k}[x_0, \ldots, x_n]/I$ and graded modules, one can pose the same question and the graded modules M with the maximal permitted cardinality of a minimal set of generators, that is those modules M such that

$$\mu(M) = e(S)\,\mathrm{rank}(M)$$

are called MGMCM modules. We are going to see that these modules are in bijection with Ulrich bundles. First of all we will see that there exists a bijection between MCM S-modules and aCM sheaves on the projective variety $\mathrm{Proj}(S)$. To explain more precisely this bijection, let us start recalling some definitions.

Definition 2.2. Let $X \subset \mathbb{P}^n$ be a variety. A coherent sheaf $\mathcal{E}$ on X is *arithmetically Cohen-Macaulay* (aCM for short) if it is locally Cohen-Macaulay (i.e., $\mathrm{depth}(\mathcal{E}_x) = \dim \mathcal{O}_{X,x}$ for any point $x \in X$) and has no intermediate cohomology:

$$\mathrm{H}^i(X, \mathcal{E}(t)) = 0 \quad \text{for all } t \in \mathbb{Z} \text{ and } 1 \leq i \leq \dim X - 1.$$

A coherent sheaf $\mathcal{E}$ on a variety $X \subset \mathbb{P}^n$ is said to be *initialized* if

$$\mathrm{H}^0(X, \mathcal{E}) \neq 0 \text{ and } \mathrm{H}^0(X, \mathcal{E}(t)) = 0 \text{ for all } t \in \mathbb{Z}_{<0}.$$

Notice that for a coherent sheaf $\mathcal{E}$ with support of positive dimension on a variety $X \subset \mathbb{P}^n$ there is always an integer t_0 such that $\mathcal{E}(t_0)$ is initialized. We usually denote it by $\mathcal{E}_{init}$.

Definition 2.3. Let $X \subset \mathbb{P}^n$ be a variety. A graded R_X-module M is a *maximal Cohen-Macaulay* module (MCM for short) if $\mathrm{depth}(M) = \dim M = \dim R_X$.

MCM modules and aCM sheaves are closely related. To state it, recall that, to any R_X-graded module E we can associate canonically a coherent sheaf $\widetilde{E}$ (see [29, page 116]). Moreover, the following holds:

Proposition 2.4. *Let $X \subset \mathbb{P}^n$ be a variety. There exists a bijection between aCM sheaves $\mathcal{E}$ on X and MCM R_X-modules given by the functors*

$$E \longrightarrow \widetilde{E} \qquad \text{and} \qquad \mathcal{E} \longrightarrow \mathrm{H}^0_*(X, \mathcal{E}) := \oplus_{t \in \mathbb{Z}} \mathrm{H}^0(X, \mathcal{E}(t)).$$

Proof. First of all, recall that, given a graded finitely generated R_X-module M there exists an exact sequence

$$0 \longrightarrow \mathrm{H}^0_{\mathfrak{m}_X}(M) \longrightarrow M \longrightarrow \mathrm{H}^0_*(X, \widetilde{M}) \longrightarrow \mathrm{H}^1_{\mathfrak{m}_X}(M) \longrightarrow 0,$$

and isomorphisms $\mathrm{H}^i_*(X, \widetilde{M}) := \oplus_{t \in \mathbb{Z}} \mathrm{H}^i(X, \widetilde{M}(t)) \cong \mathrm{H}^{i+1}_{\mathfrak{m}_X}(M)$ for $i \geq 1$, where $\mathfrak{m}_X$ denotes the irrelevant ideal of R_X.

So let $\mathcal{E}$ be an aCM sheaf on X. Then $E := \mathrm{H}^0_*(X, \mathcal{E})$ will be a finitely generated R_X-module that verifies $E \cong \mathrm{H}^0_*(X, \widetilde{E})$ and therefore $\mathrm{H}^i_{\mathfrak{m}_X}(E) = 0$ for $i \leq \dim X = \dim R_X - 1$. This allows us to conclude that E is MCM by the local cohomological criterion of depth.

On the other hand, let E be a MCM R_X-module. Then $\mathcal{E} := \widetilde{E}$ will be a locally Cohen-Macaulay sheaf and again by the previous isomorphisms we will have $\mathrm{H}^i_*(X, \mathcal{E}) = 0$ for $i = 1, \ldots, \dim X - 1$.

It was in this setting that Ulrich sheaves (without using this terminology yet) on projective hypersurfaces made the first appearance (see [6]).

Definition 2.5. Given a projective variety $X \subset \mathbb{P}^n$ embedded by the very ample line bundle $\mathcal{O}_X(1)$ and a coherent sheaf $\mathcal{E}$ on X, we say that $\mathcal{E}$ is an *Ulrich sheaf* if $\mathcal{E}$ is an initialized aCM sheaf and

$$\mathrm{h}^0(\mathcal{E}) = \deg(X) \operatorname{rank}(\mathcal{E}).$$

Remark 2.6. (1) In Proposition 2.4 we have seen that there exists a bijection between aCM sheaves $\mathcal{E}$ on a variety $X \subset \mathbb{P}^n$ and MCM R_X-modules. Under this bijection, MGMCM R_X-modules (also called linear MCM modules or Ulrich modules) correspond to Ulrich sheaves on X.

(2) The definition of Ulrich sheaf depends on the very ample line bundle $\mathcal{O}_X(1)$ of the underlying variety X and the change of the very ample line bundle can affect the existence of Ulrich sheaves.

In general, to compute $\mathrm{h}^0(X, \mathcal{E})$ for a vector bundle $\mathcal{E}$ on a variety $X \subset \mathbb{P}^n$ is not always easy. Fortunately Ulrich bundles can be characterized in many different ways.

Theorem 2.7. *Let $(X, \mathcal{O}_X(1))$ be a d-dimensional smooth variety and let $\mathcal{E}$ be an initialized vector bundle on X. The following conditions are equivalent:*

(1) $\mathcal{E}$ is Ulrich.

(2) $\mathrm{H}^i(\mathcal{E}(-t)) = 0$ for $i \geq 0$ and $1 \leq t \leq d$.

(3) $\mathrm{H}^i(\mathcal{E}(-i)) = 0$ for $i > 0$ and $\mathrm{H}^i(\mathcal{E}(-i-1)) = 0$ for $i < d$.

(4) For some (respectively all) finite linear projections $\pi : X \longrightarrow \mathbb{P}^d$, the vector bundle $\pi_*\mathcal{E}$ is the trivial vector bundle $\mathcal{O}^t_{\mathbb{P}^d}$ for some t.

(5) $\mathcal{E}$ admits a linear $\mathcal{O}_{\mathbb{P}^n}$-resolution of the form:

$$0 \longrightarrow \mathcal{O}_{\mathbb{P}^n}(-n+d)^{a_{n-d}} \longrightarrow \ldots \longrightarrow \mathcal{O}_{\mathbb{P}^n}(-1)^{a_1} \longrightarrow \mathcal{O}^{a_0}_{\mathbb{P}^n} \longrightarrow \mathcal{E} \longrightarrow 0. \tag{2.1}$$

Proof. The equivalence between (1) and (2) is given by [19, Proposition 3.2.7]. Moreover (3) is a particular case of (2).

Let us prove that (3) implies (4). We choose a finite projection $\pi : X \longrightarrow \mathbb{P}^d$. Since π is a finite morphism, $\mathrm{H}^i(X, \mathcal{E}(\ell)) = \mathrm{H}^i(\mathbb{P}^d, \pi_*(\mathcal{E}(\ell)))$ for all i and ℓ. Therefore, $\pi_*(\mathcal{E})$ satisfies the same vanishing conditions as $\mathcal{E}$. Let us quickly see that a coherent sheaf $\mathcal{F}$ on $\mathbb{P}^d$ with the vanishing conditions (3) should be of the form $\mathcal{O}_{\mathbb{P}^d}^t$ for some $t \in \mathbb{Z}$. First of all, $\mathcal{F}$ is 0-regular and therefore $\mathrm{H}^i(\mathcal{F}(t_i)) = 0$ for $i > 0$ and $t_i \geq -i$. To conclude, we are going to see by induction on d that $\mathcal{F}$ is an aCM bundle and, therefore, by Horrocks Theorem (see [34]), of the desired form. The case $d = 1$ is immediate by Grothendieck's Theorem. So let us suppose that it is true for $d - 1$ and consider the short exact sequence

$$0 \longrightarrow \mathcal{F}(-1) \longrightarrow \mathcal{F} \longrightarrow \mathcal{F}_{|H} \longrightarrow 0$$

for a generic hyperplane section $H \cong \mathbb{P}^{d-1}$. Considering the long exact sequence associated to it we see that $\mathcal{F}_{|H}$ satisfies the same vanishing conditions of the statement for $d-1$. Therefore, by induction hypothesis, $\mathcal{F}_{|H}$ splits as a direct sum of trivial line bundles $\mathcal{O}_{\mathbb{P}^{d-1}}$. In particular, we have injections $\mathrm{H}^i(\mathcal{F}(t-1)) \hookrightarrow \mathrm{H}^i(\mathcal{F}(t))$ for $1 \leq i \leq d-1$ and $t \leq -1$. Therefore $\mathcal{F}$ is an aCM bundle and we are done.

Let us see that (4) implies (5). The hypothesis $\pi_*\mathcal{E} \cong \mathcal{O}_{\mathbb{P}^d}^t$ for some t implies that $M = \mathrm{H}^0_*(\mathcal{E})$ is a free module over $\mathbf{k}[x_0, \cdots, x_d] = \mathrm{H}^0_*(\mathcal{O}_{\mathbb{P}^d})$ generated in degree 0. This means that M is a MCM of dimension $d+1$ over R with a linear resolution:

$$0 \longrightarrow R(d-n)^{a_{n-d}} \longrightarrow \cdots \longrightarrow R(-1)^{a_1} \longrightarrow R^{a_0} \longrightarrow M \longrightarrow 0.$$

Sheafifying this last resolution we get (2.1).

Finally we will prove that (5) implies (2). Cutting the linear resolution (2.1) into short exact sequences, taking cohomology and using the fact that $\mathrm{H}^i(\mathbb{P}^d, \mathcal{O}_{\mathbb{P}^d}(-t)) = 0$ for all $i \geq 0$ and $1 \leq t \leq d$, we get $\mathrm{H}^i(\mathcal{E}(-t)) = 0$ for $i \geq 0$ and $1 \leq t \leq d$ which proves what we want.

The existence of Ulrich bundles on a given variety is, in general, a nontrivial problem. Let us start with examples of Ulrich line bundles.

Example 2.8. (1) $\mathcal{O}_{\mathbb{P}^n}$ is an Ulrich bundle on $\mathbb{P}^n$ and, by Horrocks Theorem (see [34]), it is the only indecomposable Ulrich bundle on $\mathbb{P}^n$.

(2) Projective varieties $X \subset \mathbb{P}^n$ with $\mathrm{Pic}(X) \cong \mathbb{Z}$ do not carry Ulrich line bundles except when X is a linear space. The result is no longer true for varieties $X \subset \mathbb{P}^n$ with $\mathrm{Pic}(X) \cong \mathbb{Z}^\rho$, $\rho > 1$. For instance, the quadric $X \cong \mathbb{P}^1 \times \mathbb{P}^1 \subset \mathbb{P}^3$ supports two Ulrich line bundles: $\mathcal{O}_{\mathbb{P}^1 \times \mathbb{P}^1}(1,0)$ and $\mathcal{O}_{\mathbb{P}^1 \times \mathbb{P}^1}(0,1)$. More generally, a smooth cubic surface $S \subset \mathbb{P}^3$ carries a finite number of Ulrich line bundles: $\mathcal{O}_S(D)$ being $D \subset S$ a rational normal cubic curve.

(3) We consider the Segre variety $\mathbb{P}^n \times \mathbb{P}^m$ embedded in $\mathbb{P}^{nm+n+m}$ by the very ample line bundle $\mathcal{O}_{\mathbb{P}^n \times \mathbb{P}^m}(1,1)$. Using the *Künneth formula* we can prove that a line bundle $\mathcal{O}_{\mathbb{P}^n \times \mathbb{P}^m}(a,b)$ on a Segre variety $\mathbb{P}^n \times \mathbb{P}^m \subset \mathbb{P}^{nm+n+m}$ is an

Ulrich line bundle if and only if $(a, b) \in \{(m, 0), (0, n)\}$. If we change the very ample line bundle and for example we fix as very ample line bundle $\mathcal{O}_{\mathbb{P}^n \times \mathbb{P}^m}(2, 3)$ we check, using again Künneth formula that there is no Ulrich line bundle on $\mathbb{P}^n \times \mathbb{P}^m$ with respect to the very ample line bundle $\mathcal{O}_{\mathbb{P}^n \times \mathbb{P}^m}(2, 3)$.

Since the entrance of Ulrich sheaves in the realm of algebraic geometry, a leading question has been the existence of Ulrich bundles on a given projective variety. This problem is far from being solved. However, in the particular case of curves, the existence of Ulrich bundles can be settled.

Indecomposable Ulrich bundles of higher rank, by Grothendieck's Theorem, can not exist on rational curves. Vector bundles on elliptic curves where classified by Atiyah in [3]. Using this classification, the following result can be obtained:

Theorem 2.9. *Let $(C, \mathcal{O}_C(1))$ be a smooth elliptic curve. For any $r \geq 1$, there exists a one dimensional family of indecomposable rank r Ulrich bundles on C. They are strictly semistable.*

Proof. See [19, Theorem 3.4.6].

Finally, for smooth curves $C \subset \mathbb{P}^n$ of genus $g > 1$ we have:

Theorem 2.10. *Let $(C, \mathcal{O}_C(1))$ be a smooth curve of genus $g > 1$ and degree k and take $H \in |\mathcal{O}_C(1)|$. Then C supports stable rank r Ulrich bundles and they form an open subset of the moduli space $M_{C,H}(r; r(k + g - 1))$ of stable vector bundles of rank r and degree $r(k + g - 1)$.*

Proof. See [19, Theorem 3.4.7].

Although the existence of Ulrich bundles on arbitrary curves is known, things become much more involved when we move forward to higher dimension. Indeed, already in the case of smooth projective surfaces, the existence of Ulrich bundles is only known in some cases, as they can be surfaces $S \subset \mathbb{P}^3$, surfaces of minimal degree, del Pezzo surfaces, etc. However, for instance in the case of surfaces of general type, very sparse results have been obtained by now. There are scattered examples of Ulrich bundles on higher dimensional varieties as Grassmannnians, some flag varieties, hypersurfaces in $\mathbb{P}^n$ among others.

All these results support and can be seen as a partial answer to the following conjecture due to Eisenbud and Schreyer ([25]).

Conjecture 2.11. Any smooth projective variety $(X, \mathcal{O}_X(1))$ supports an Ulrich bundle.

3 A Geometrical Classical Problem

In spite Ulrich bundles have they roots in the realm of commutative algebra, they move very soon to a more geometrical side when, more than a century ago, geometers seek for determinantal representations of hypersurfaces. More precisely, they were focused on the following:

Problem 3.1. *Given a homogeneous polynomial f of degree d, find a square matrix $M = (\ell_{ij})$ of size d with linear entries such that*

$$f = \det(M).$$

The interest in determinantal representations of homogeneous polynomials raised during the nineteenth century. It seems that, the problem of representing a homogenous form as the determinant of a matrix of linear forms was first clearly stated by Hesse in [31]. In addition, in [32] he was able to prove that the smooth quartic plane curve has a symmetric determinantal representation. Hesse also proved that for smooth plane cubic curves the representation follows from the fact that any smooth plane cubic curve can be written in three ways as the Hessian curve. The cases of cubic and quartic surfaces were already solved by Grassmann in [28] and Cremona in [21], respectively.

Later on, Cayley approached the symmetric representational case for these degrees in [13]. It was Coble [15] the first to understand when a quartic surface S_4 can be represented as a determinant of a matrix of linear forms. It turns out that S_4 can be defined by the determinant of a matrix with linear entries if and only if it contains an aCM curve of degree 6 and genus 3. By Noether-Lefschetz Theorem, a very general surface $S \subset \mathbb{P}^3$ of degree $d \geq 4$ has Picard group generated by the hyperplane section and therefore a very general quartic surface does not contain those kind of curves. Other examples of curves and surfaces are treated by Schur in [36]. Already in [22], it was realised that only a general plane curve of arbitrary degree, quadric surfaces and cubic surfaces can be represented by a determinant of linear forms. Their existence for smooth plane curves of arbitrary degree was given in [23]. Therefore, people realised that would be worth of finding determinantal representation for powers f^r, $r \geq 2$, of the form f defining an integral surface in $\mathbb{P}^3$, or more generally:

Problem 3.2. *Find determinantal representation for powers f^r, $r \geq 2$, of the form f defining an integral hypersurface in $\mathbb{P}^n$, $n \geq 3$.*

The first case would be $r = 2$ and adding the condition that the matrix M is skew-symmetric ($M^t = -M$). $\det(M)$ is a perfect square of an element known as the pfaffian $\mathrm{Pf}(M)$ of M and $f = \mathrm{Pf}(M)$. This situation has been less studied in the literature. In [1] it is proved that a generic cubic threefold in $\mathbb{P}^4$ can be written as a linear pfaffian. Beauville (see [6]) studied systematically which hypersurfaces have a pfaffian representation.

The main goal of this section is to explain how the existence of a rank r Ulrich bundle on a hypersurface $X \subset \mathbb{P}^n$ defined by a homogeneous form f of degree d, implies that f^r can be written as the determinant of a square matrix of size dr with linear entries (see, for instance, [16,22,28,36]), and viceversa. As a consequence of this relation we will see that for any hypersurface X defined by an homogeneous form f, there exists an integer $r \geq 1$ such that f^r has a linear determinantal representation (Proposition 3.6).

Definition 3.3. Let $0 \neq f \in R = \mathbf{k}[x_0, \cdots, x_n]$ be a homogeneous polynomial of degree $d \geq 2$. A *linear matrix factorization* of f of size m is an equation

$$fI_m = \alpha_0\alpha_1 \cdots \alpha_{d-1}$$

where the α_i are square matrices of size m whose entries are linear forms in R and I_m stands for the identity matrix of size m.

Using the fact that if $0 \neq f, g \in \mathbf{k}[x_0, \cdots, x_n]$ are homogeneous polynomials of degree d with matrix factorizations of size m and l respectively, then $h := f+g$ has a linear matrix factorization, of size dlm we obtain:

Corollary 3.4. *Let $0 \neq f \in \mathbf{k}[x_0, \cdots, x_n]$ be a homogeneous polynomial of degree d with s summands of products of linear forms. Then, f has a linear matrix factorization of size $m = d^{s-1}$.*

The relation with Ulrich bundles is stated in the following result:

Proposition 3.5. *Let $X \subset \mathbb{P}^n$ be an integral hypersurface of degree d defined by a homogeneous polynomial f with a linear matrix factorization of size m. Then, X supports an Ulrich bundle, of rank $\frac{m}{d}$.*

Proof. By assumption f has a linear matrix factorization $fI_m = \alpha_0\alpha_1 \cdots \alpha_{d-1}$ of size m. Denote by $\psi = \alpha_0\alpha_1 \cdots \alpha_{d-2}$ and by $\varphi = \alpha_{d-1}$. Then

$$fI_m = \psi\varphi$$

is a two-factor matrix factorization of f. Hence, by Corollary [19, Corollary 2.5.13] there is an aCM bundle $\mathcal{E} = \operatorname{coker}(\varphi)$ on X given by the exact sequence:

$$0 \longrightarrow \mathcal{O}_{\mathbb{P}^n}^m(-1) \xrightarrow{\varphi} \mathcal{O}_{\mathbb{P}^n}^m \longrightarrow \mathcal{E} \longrightarrow 0.$$

In particular,

$$\mathrm{H}^0(X, \mathcal{E}(-1)) = 0 \quad \text{and} \quad \mathrm{H}^{n-1}(X, \mathcal{E}(1-n)) = 0.$$

Thus, by Proposition 2.7, $\mathcal{E}$ is an Ulrich bundle on X. Finally, it follows from the above exact sequence that

$$m = \mathrm{h}^0(\mathcal{E}) = \deg(X)\operatorname{rank}(\mathcal{E})$$

from which we deduce that $\operatorname{rank}(\mathcal{E}) = \frac{m}{d}$.

Proposition 3.6. *Let $0 \neq f \in R = \mathbf{k}[x_0, \cdots, x_n]$ be a homogeneous polynomial of degree d defining an integral hypersurface $X \subset \mathbb{P}^n$. Denote by s the number of summands of products of linear forms of f. Then, there exists a matrix M of linear forms such that*

$$\det(M) = f^r$$

where $r = d^{s-2}$.

Proof. By Corollary 3.4 and Proposition 3.5, X supports an Ulrich bundle $\mathcal{E}$ of rank d^{s-2}. By Theorem 2.7, to say that $\mathcal{E}$ is Ulrich means that its minimal free resolution has the following shape

$$0 \longrightarrow \mathcal{O}^m_{\mathbb{P}^n}(-1) \xrightarrow{M} \mathcal{O}^m_{\mathbb{P}^n} \longrightarrow \mathcal{E} \longrightarrow 0,$$

where M is a square matrix of linear forms. Since $\mathcal{E}$ is annihilated by f, the multiplication by f on $\mathcal{O}^m_{\mathbb{P}^n}$ factors through M. So, we can find a matrix N of degree $d-1$ forms such that $MN = fI_m$. In particular, $\det(M)\det(N) = f^m$. Thus, if f is irreducible the determinant of M is a power f^r of f. Finally, by [24, Proposition 5.6], r is the rank of $\mathcal{E}$.

We have seen the connection between linear matrix factorizations of a given hypersurface $X \subset \mathbb{P}^n$ and Ulrich bundles supported on $(X, \mathcal{O}_X(1))$. Indeed, this connection can be introduced in full generality for an arbitrary projective variety $(X, \mathcal{O}_X(1))$ in terms of its associated *Cayley-Chow form.* Cayley-Chow forms were introduced for curves in the projective space $\mathbb{P}^3$ in [11,12]. Later on, the concept was generalized to arbitrary varieties by Bertini, Chow and van der Warden (see [38]). Let us end this section introducing this concept and its relation with Ulrich bundles.

Let $(X, \mathcal{O}_X(1))$ be a d-dimensional projective variety providing an embedding $X \subset \mathbb{P}^n$. Let us consider the following incidence relation

$$\begin{array}{ccc} & \mathbb{F} := \{(x,H) \mid x \in H\} \subset \mathbb{P}^n \times \mathrm{Gr}(n-d-1,n) & \\ \swarrow^{p} & & \searrow^{q} \\ \mathbb{P}^n & & \mathrm{Gr}(n-d-1,n) \end{array}$$

where we have denoted by p and q the corresponding natural projections.

Definition 3.7. Let $(X, \mathcal{O}_X(1))$ be a d-dimensional projective variety embedded in $\mathbb{P}^n$. With the above notations, the variety

$$D_X := q \circ p^{-1}(X) = \{H \in \mathrm{Gr}(n-d-1,n) \mid H \cap X \neq \emptyset\} \subset \mathrm{Gr}(n-d-1,n)$$

has codimension one in $\mathrm{Gr}(n-d-1,n)$ which means that it is a divisor. D_X is called the *Cayley-Chow divisor* of X.

Therefore D_X is given by a single equation f_X in the Plücker embedding. In fact, $\mathrm{Pic}(\mathrm{Gr}(n-d-1,n)) \cong \mathbb{Z} < \mathcal{O}_{\mathrm{Gr}(n-d-1,n)}(1) >$ and $\mathcal{O}_{\mathrm{Gr}(n-d-1,n)}(1)$ gives the *Plücker embedding* $\mathrm{Gr}(n-d-1,n) \subset \mathbb{P}(V)$, $V = \wedge^{n-d}\mathbf{k}^{n+1}$. Moreover, this is a projectively normal embedding and therefore $D_X \subset \mathrm{Gr}(n-d-1,n)$ is cut out in $\mathrm{Gr}(n-d-1,n)$ by a single homogeneous polynomial f_X in the *Plücker coordinates.*

Definition 3.8. Let $(X, \mathcal{O}_X(1))$ be a d-dimensional projective variety embedded in $\mathbb{P}^n$. With the above notations, the homogeneous form f_X in the Plücker coordinates associated to the variety X is called the *Cayley-Chow form* of X.

An important feature of Cayley-Chow forms f_X is that one can recover the original variety X from them. Set-theoretically, this result was proved by Van der Warden in [38]. Scheme-theoretically, we have the following result:

Theorem 3.9. *Let $X \subset \mathbb{P}^n$ be a hypersurface or a smooth variety. Then the Cayley-Chow form f_X of X determines completely the variety X.*

Proof. See [9, Theorem 1.14] and [25, Proposition 1.3].

In general, it is an intriguing question to find the Cayley-Chow form f_X of a given variety X. Nevertheless, the Cayley-Chow form f_X of a projective variety $(X, \mathcal{O}_X(1))$ has a very pleasant presentation when $(X, \mathcal{O}_X(1))$ supports an Ulrich bundle $\mathcal{E}$.

In fact, let M be the graded module of twisted global sections over the polynomial ring $R = \mathbf{k}[x_0, \cdots, x_n] = \mathrm{H}^0_*(\mathcal{O}_{\mathbb{P}^n})$. In the proof of Theorem 2.7 we have seen that M has a a linear resolution:

$$0 \longrightarrow R(c)^{a_c} \longrightarrow \cdots \longrightarrow R(-1)^{a_1} \longrightarrow R^{a_0} \longrightarrow M \longrightarrow 0. \tag{3.1}$$

Denote by W^* the vector space generated by the variables $x_0, \cdots, x_n$. Notice that the maps d_i in the resolution (3.1) can be represented by matrices D_i of size $a_i \times a_{i+1}$ with entries in W^*. Since we have a complex, the product $D_{i-1}D_i$ is the zero matrix while doing this product, the entries are elements in the ring $R = Sym(W^*)$. Let us now denote by B the exterior algebra on W^* and consider the entries of each D_i in B. Then, while considering the product

$$D = D_0 D_1 \cdots D_c$$

the multiplications of the entries are performed in B and c is the codimension of the variety X which supports the Ulrich bundle $\mathcal{E}$. If r is the rank of $\mathcal{E}$ and d the degree of X, then D is a $dr \times dr$ matrix with entries in $\wedge^c W^*$.

It can be seen that $\det(D)$ is a form of degree rd in $\mathrm{Sym}(\wedge^c W^*)$ which equals the $r-th$ power of the Cayley-Chow form f_X of X. In summary, keeping this notation we have:

Theorem 3.10. *Let $(X, \mathcal{O}_X(1))$ be a variety and let $\mathcal{F}$ be an Ulrich bundle on it. Then, the Cayley-Chow form f_X of X satisfies*

$$f_X^r = f_{\mathcal{F}} = \det(D).$$

4 Ulrich Bundles and Mathematical Physics

The notion of an instanton comes from mathematical physics. It denotes a solution of an equation of motion in classical field theory which describes a particle localized both in space and in time. Mathematically, it is a self-dual connection on a principal bundle on a four-dimensional Riemannian manifold. The Penrose-Ward correspondence identifies instantons on the four-dimensional sphere S^4 with holomorphic instanton bundles on the three-dimensional complex projective space $\mathbb{P}^3$. This construction turned out to be a major motivation for studying vector bundles on complex projective spaces and similar algebraic varieties.

We have seen that, on the contrary of instanton bundles, Ulrich bundles were born in the context of commutative algebra and they are related with the so called maximally generated maximal Cohen Macaulay modules. In this section we will relate the notion of Ulrich bundle born in the context of commutative algebra with the notion of instanton bundle born in the context of mathematical physics. In fact, we will relate the existence of Ulrich bundles on the d-uple Veronese embedding $(\mathbb{P}^3, \mathcal{O}_{\mathbb{P}^3}(d))$, $d \geq 2$, of $\mathbb{P}^3$ with the existence of instanton bundles on $\mathbb{P}^3$.

To this end, let us start gathering together the definition and the results on instanton bundles that we will use to achieve this relation. We start recalling the definition of monad.

Definition 4.1. A *monad* over $\mathbb{P}^n$ is a complex of vector bundles:

$$M_\bullet : \quad 0 \longrightarrow \mathcal{A} \xrightarrow{\alpha} \mathcal{B} \xrightarrow{\beta} \mathcal{C} \longrightarrow 0$$

with $\beta\alpha = 0$, $\alpha : \mathcal{A} \longrightarrow \mathcal{B}$ an injective bundle-map and $\beta : \mathcal{B} \longrightarrow \mathcal{C}$ surjective. The vector bundle $\mathcal{E} := \ker(\beta)/\operatorname{im}(\alpha)$ is called the cohomology bundle of the monad.

A monad $0 \longrightarrow \mathcal{A} \xrightarrow{\alpha} \mathcal{B} \xrightarrow{\beta} \mathcal{C} \longrightarrow 0$ has a so called display. This is a commutative diagram with exact rows and columns:

$$\begin{array}{ccccccccc}
 & & & & 0 & & 0 & & \\
 & & & & \downarrow & & \downarrow & & \\
0 & \longrightarrow & \mathcal{A} & \longrightarrow & \mathcal{K} & \longrightarrow & \mathcal{E} & \longrightarrow & 0 \\
 & & \| & & \downarrow & & \downarrow & & \\
0 & \longrightarrow & \mathcal{A} & \xrightarrow{\alpha} & \mathcal{B} & \longrightarrow & \mathcal{Q} & \longrightarrow & 0 \\
 & & & & \downarrow & & \downarrow & & \\
 & & & & \mathcal{C} & = & \mathcal{C} & & \\
 & & & & \downarrow & & \downarrow & & \\
 & & & & 0 & & 0 & &
\end{array}$$

where $\mathcal{K} := \ker(\beta)$ and $\mathcal{Q} := \operatorname{Coker}(\alpha)$. From the display one easily deduces that if a vector bundle $\mathcal{E}$ on $\mathbb{P}^n$ is the cohomology bundle of a monad $M_\bullet$, then

(i) $\operatorname{rk}(\mathcal{E}) = \operatorname{rk}(\mathcal{B}) - \operatorname{rk}(\mathcal{A}) - \operatorname{rk}(\mathcal{C})$, and
(ii) $c_t(\mathcal{E}) = c_t(\mathcal{B})c_t(\mathcal{A})^{-1}c_t(\mathcal{C})^{-1}$.

As we quoted before, the notion of an instanton comes from mathematical physics. It denotes a solution of an equation of motion in quantum theory which describes a (pseudo) particle which exists at only one moment in time. Mathematically, it is a self-dual connection on a principal bundle on a four-dimensional Riemannian manifold. The Penrose- Ward transform turns instantons on the four-dimensional sphere S^4 into holomorphic instanton bundles on the three-dimensional complex projective space $\mathbb{P}^3$ [4,5].

Instanton bundles $\mathcal{E}$ on a three dimensional projective space $\mathbb{P}^3$ were originally defined as the algebraic counterpart that permitted to solve a crucial problem in theoretical physics, namely the classification of the minimum action solutions of $SU(2)$ Yang-Mills fields in Euclidean 4-space. Let us recall the original definition.

Definition 4.2. A *mathematical instanton bundle* on $\mathbb{P}^3$ with quantum number k is a rank 2 stable vector bundle $\mathcal{E}$ on $\mathbb{P}^3$ with Chern classes $c_1(\mathcal{E}) = 0$, $c_2(\mathcal{E}) = k$ and verifying the instanton condition $\mathrm{H}^2(\mathcal{E}(-2)) = 0$.

The Penrose–Ward correspondence also exists on higher odd-dimensional projective spaces. In fact, Salamon, Corrigan, Goddard, and Kent introduced a correspondence between self-dual connections on $\mathrm{Sp}_n(\mathbb{C})$-bundles on the n-dimensional quaternionic projective space $\mathbb{HP}^n$ and symplectic instanton bundles on the complex projective space $\mathbb{P}^{2n+1}$, $n \geq 1$ [14,20,35]. Later on the definition was also generalized to vector bundles of higher rank.

Definition 4.3. A *mathematical instanton bundle* on $\mathbb{P}^3$ with quantum number k is a rank r vector bundle $\mathcal{E}$ on $\mathbb{P}^3$ with Chern classes $c_1(\mathcal{E}) = 0$, $c_2(\mathcal{E}) = k$ and verifying the instanton conditions $\mathrm{H}^0(\mathcal{E}(-1)) = \mathrm{H}^3(\mathcal{E}(-3)) = \mathrm{H}^1(\mathcal{E}(-2)) = \mathrm{H}^2(\mathcal{E}(-2)) = 0$.

As an easy consequence of the above definition and applying Beilinson's spectral sequence, we get any rank r instanton bundle on $\mathbb{P}^3$ as the cohomology bundle of a linear monad. Indeed, we have the following result.

Proposition 4.4. *Let $\mathcal{E}$ be a rank r instanton bundle on $\mathbb{P}^3$ with quantum number k. Then, $\mathcal{E}$ is the cohomology bundle of a linear monad of the following type:*

$$0 \longrightarrow \mathcal{O}(-1)^k \xrightarrow{\alpha} \mathcal{O}^{2k+r} \xrightarrow{\beta} \mathcal{O}(1)^k \longrightarrow 0 \tag{4.1}$$

and, vice versa, any vector bundle $\mathcal{E}$ on $\mathbb{P}^3$ which appears as the cohomology bundle of a monad of the type (4.1) is a rank r instanton bundle on $\mathbb{P}^3$ with quantum number k.

Now we are ready to state the relation between instanton bundles and Ulrich bundles. To this end, for any positive integer d, we denote by $(\mathbb{P}^3, \mathcal{O}(d))$ the 3-dimensional *Veronese variety* defined as the image of the Veronese embedding $\mathbb{P}^3 \longrightarrow \mathbb{P}^{\binom{3+d}{d}-1}$ given by the very ample line bundle $\mathcal{O}_{\mathbb{P}^3}(d)$.

Theorem 4.5. *There is a one to one correspondence between rank r initialized Ulrich bundles $\mathcal{E}$ on $(\mathbb{P}^3, \mathcal{O}(d))$ and rank r instanton bundles $\mathcal{F}$ on $\mathbb{P}^3$ with quantum number $\frac{r(d^2-1)}{6}$ and natural cohomology.*

Proof. We will prove that if $\mathcal{E}$ is a rank r initialized Ulrich bundle on $(\mathbb{P}^3, \mathcal{O}(d))$ then $\mathcal{E}(2-2d)$ is a rank r instanton bundle on $\mathbb{P}^3$ with quantum number $\frac{r(d^2-1)}{6}$ and natural cohomology; and vice versa if $\mathcal{F}$ is a rank r instanton bundle on $\mathbb{P}^3$ with quantum number $\frac{r(d^2-1)}{6}$ and natural cohomology then $\mathcal{F}(2d-2)$ is a rank r initialized Ulrich bundle on $(\mathbb{P}^3, \mathcal{O}(d))$.

Assume that $\mathcal{E}$ is a rank r initialized Ulrich bundle on $(\mathbb{P}^3, \mathcal{O}(d))$.
Claim: $\mathcal{E}$ has natural cohomology and

$$\chi\mathcal{E}(t) = \frac{r}{6}(t+d)(t+2d)(t+3d).$$

Proof of the Claim: Since $\mathcal{E}$ is a rank r initialized Ulrich bundle on $(\mathbb{P}^3, \mathcal{O}(d))$, we have $\mathrm{H}^i(\mathcal{E}(-td)) = 0$ for $1 \le t \le 3$. Therefore,

$$\chi\mathcal{E}(t) = a(t+d)(t+2d)(t+3d)$$

with $a \in \mathbb{Q}$. To determine a we use again that $\mathcal{E}$ is a rank r initialized Ulrich bundle on $(\mathbb{P}^3, \mathcal{O}(d))$. So, we have $\mathrm{H}^i\,\mathcal{E} = 0$ for $i = 1,2,3$ and $\mathrm{h}^0\,\mathcal{E} = \deg((\mathbb{P}^3, \mathcal{O}(d)))\mathrm{rk}\mathcal{E} = rd^3$ and we conclude that $a = \frac{r}{6}$ which proves what we want.

By the Claim and [25, Theorem 5.1], $\mathcal{E}$ has natural cohomology, $\chi\mathcal{E}(t) = \frac{r}{6}(t+d)(t+2d)(t+3d)$ and

$$\mathrm{H}^i(\mathcal{E}(t)) \neq 0 \Leftrightarrow \begin{cases} i=0 & -d<t \\ i=1 & -2d<t<-d \\ i=2 & -3d<t<-2d \\ i=3 & t<-3d. \end{cases}$$

Therefore, $\mathcal{E}(2-2d)$ is a rank r vector bundle on $\mathbb{P}^3$ with Chern classes $c_1(\mathcal{E}(2-2d)) = 0$ and $c_2(\mathcal{E}(2-2d)) = \frac{1}{6}r(d^2-1)$. It has natural cohomology and it satisfies the instanton conditions

$$\mathrm{H}^0(\mathcal{E}(2-2d)(-1)) = \mathrm{H}^1(\mathcal{E}(2-2d)(-2)) = 0$$

$$\mathrm{H}^2(\mathcal{E}(2-2d)(-2)) = \mathrm{H}^3(\mathcal{E}(2-2d)(-3)) = 0$$

which proves what we want.

Let $\mathcal{F}$ be a rank r instanton bundle on $\mathbb{P}^3$ with quantum number $\frac{r(d^2-1)}{6}$ and natural cohomology. Since $\mathrm{rk}(\mathcal{F}) = r$, $c_1(\mathcal{F}) = c_3(\mathcal{F}) = 0$ and $c_2(\mathcal{F}) = \frac{1}{6}r(d^2-1)$, applying Riemann-Roch theorem we obtain:

$$\chi\mathcal{F}(2d-2)(t) = \frac{r}{6}(t+d)(t+2d)(t+3d).$$

This last equality together with the hypothesis that $\mathcal{F}$ has natural cohomology implies that

$$\mathrm{H}^i(\mathbb{P}^3, \mathcal{F}(2d-2)(-td)) = 0$$

for all $i \ge 0$ and $1 \le t \le 3$. Thus, $\mathcal{F}(2d-2)$ is a rank r initialized Ulrich bundle on $(\mathbb{P}^3, \mathcal{O}(d))$.

This relation between instanton bundles and Ulrich bundles pushed the authors in [18] to state the following conjecture:

Conjecture 4.6. A rank r Ulrich bundle on $(\mathbb{P}^3, \mathcal{O}(d))$ exists if and only if $r(d^2-1) = 0$ (mod. 6).

Notice that if there exists a rank r Ulrich bundle on $(\mathbb{P}^3, \mathcal{O}(d))$, then there exsits an instanton bundle on $\mathbb{P}^3$ with quantum number $\frac{r}{6}(d^2-1) \in \mathbb{Z}$ and hence $r(d^2-1) = 0$ (mod. 6). The knowledge of instanton bundles has allowed to prove in [18] the converse of Conjecture 4.6 for the following values of d and r:

- $d = 2^\alpha$ and any r,
- $d = 3$ and any r,
- $d = 6n$, $3 \nmid n$ and any r, and
- $d = 2n$, $3 \nmid n$ and any r.

5 Ulrich Bundles and Computer Vision

In this section we will see how the existence of certain Ulrich bundles can be used to solve tiny reconstruction problems in computer vision. The content of this section is part of the work [27], by Floystad, Kileel and Ottaviani, where all the details can be found. From now on, we will assume that $\mathbf{k} = \mathbb{C}$. Let us start introducing what is called the *essential variety*. Let $\mathcal{V} \in \mathbb{R}^{3\times 3}$ be the essential variety defined by the conditions

$$\mathcal{V} = \{M \in \mathbb{R}^{3\times 3} | \det(M) = 0, 2(MM^T)M - tr(MM^T)M = 0\}. \tag{5.1}$$

These ten cubics minimally generate the real radical ideal of $\mathcal{V}$. We denote by $\mathcal{V}_{\mathbb{C}}$ the projective variety in $\mathbb{P}^8$ given by the complex solution of (5.1). In [27] the authors proved that it has codimension 3 and degree 10 and that it is isomorphic to a hyperplane section of the variety $Y_{4,2}$ of complex symmetric 4×4 matrices of rank ≤ 2.

The next goal is to construct the Cayley-Chow form of the essential variety. As we already have seen, this can be achieved provided we have an Ulrich bundle on this variety. To this end, in [27] the authors prove the existence of two Ulrich bundles on $Y_{4,2}$ that restrict to two Ulrich bundles $\mathcal{E}_1$ and $\mathcal{E}_2$ on the essential variety $\mathcal{V}_{\mathbb{C}}$. As explained in Sect. 2, the matrices appearing in the resolution of $\mathcal{E}_1$ and $\mathcal{E}_2$ respectively give rise to two 20×20 skew-symmetric matrices M_1 and M_2 (explicitly available in http://math.berkeley.edu/jkileel/ChowFormulas.html) whose Pfaffians equal the Cayley-Chow form of $\mathcal{V}_{\mathbb{C}}$. These Cayley-Chow forms of $\mathcal{V}_{\mathbb{C}}$ are elements in the homogeneous coordinate ring of the Grassmannian $\mathrm{Gr}(\mathbb{P}^2, \mathbb{P}^8)$. This ring can be seen as the polynomial ring in 84 (dual) Plucker variables modulo 1050 Plucker quadrics.

Let us now see which role play these explicit expressions of the Cayley-Chow form of the essential variety in computer vision. In fact these expressions provide an important tool for the problem of detecting if a set of image point correspondences $\{(x^{(i)}, y^{(i)}) \in \mathbb{R}^2 \times \mathbb{R}^2 | 1 \leq i \leq m\}$ comes from m world points in $\mathbb{R}^3$ and two calibrated cameras. It furnishes an exact solution for $m = 6$ and it behaves well given noisy input. Mathematically, we can consider the system of equations:

$$\begin{cases} A\widetilde{X^{(i)}} \equiv \widetilde{x^{(i)}} \\ B\widetilde{X^{(i)}} \equiv \widetilde{y^{(i)}}. \end{cases} \tag{5.2}$$

Here $\widetilde{x^{(i)}} \in \mathbb{P}^2$ and $\widetilde{y^{(i)}} \in \mathbb{P}^2$ are the given image points. The unknowns are two 3×4 matrices A, B with rotations in their left 3×3 blocks and $m = 6$ points $\widetilde{X^{(i)}} \in \mathbb{P}^3$. These represent calibrated cameras and world points, respectively.

Taking either of the 20×20 skew-symmetric matrices M_1 or M_2, the authors in [27] proved:

Theorem 5.1. *There exists an explicit* 20×20 *skew-symmetric matrix* $\mathcal{M}(x, y)$ *of degree* $\leq (6, 6)$ *polynomials over* $\mathbb{Z}$ *in the coordinates of* $\{(x^{(i)}, y^{(i)})\}$ *with the following properties. If (5.2) admits a complex solution then* $\mathcal{M}(x^{(i)}, y^{(i)})$ *is rank-deficient. Conversely, the variety of point correspondences* $\{(x^{(i)}, y^{(i)})\}$ *such that* $\mathcal{M}(x, y)$ *is rank-deficient contains a dense subset for which (5.2) admits a complex solution.*

From the point of view of computer vision, this result contributes a complete characterization for an "almost-minimal" problem. Here the motivation is *3D reconstruction*. Given multiple images of a world scene, taken by cameras in an unknown configuration, we want to estimate the camera configuration and a 3D model of the world scene. Algorithms for this are complex and successful. For instance there are reconstructions from 150.000 images. By contrast, the system (5.2) encodes a tiny reconstruction problem. Suppose we are given 6 point correspondences in two calibrated pictures (the right hand side in the system). We wish to reconstruct both the two cameras and the six world points (left hand side of the system). If an exact solution exists then it is typically unique, modulo the natural symmetries. However, an exact solution does not always exists. In order to this to happen, a giant polynomial of degree 120 in the 24 variables on the right-hand side has to vanish. Theorem 5.1 gives an explicit matrix formula for that polynomial.

Acknowledgements. The author was partially supported by PID2019-104844GB-I00

References

1. Adler, A., Ramanan, S.: Moduli of Abelian Varieties. Lecture Notes in Mathematics, vol. 1644. Springer, Heidelberg (1996). https://doi.org/10.1007/BFb0093659
2. Alexander, J., Hirschowitz, A.: Polynomial interpolation in several variables. J. Alg. Geom. **4**, 201–22 (1995)
3. Atiyah, M.F.: Vector bundles over an elliptic curve. Proc. Lond. Math. Soc. **7**, 414–452 (1957)

4. Atiyah, M.F.: Geometry on Yang-Mills fields, p. 99. Scuola Normale Superiore Pisa, Pisa (1979)
5. Atiyah, M.F., Ward, R.S.: Instantons and algebraic geometry. Comm. Math. Phys. **55**, 117–124 (1977)
6. Beauville, A.: Determinantal hypersurfaces. Michigan Math. J. **48**, 39–64 (2000)
7. Beauville, A.: An Introduction to Ulrich Bundles. Eur. J. Math. (2017). https://doi.org/10.1007/s40879-017-0154-4
8. Brennan, J.P., Herzog, J., Ulrich, B.: Maximally generated Cohen-Macaulay modules. Math. Scandinavica **61**, 181–203 (1987)
9. Catanese, F.: Chow varieties, Hilbert schemes and moduli spaces of surfaces of general type. J. Algebraic Geom. **1**(4), 561–595 (1992)
10. Catanese, F.: Cayley forms and self-dual varieties. Proc. Edinbourgh Math. Soc. **57**(1), 89–109 (2014)
11. Cayley, A.: On a new analytical representation of curves in space. Q. J. Pure Appl. Math. **III**, 225–236 (1860)
12. Cayley, A.: On a new analytical representation of curves in space. Q. J. Pure Appl. Math. **V**, 81–86 (1862)
13. Cayley, A.: A Memoir on Quartic Surfaces. Proc. Lond. Math. Soc. **3**, 19–69 (1869-71)
14. Capria, M.M., Salamon, S.M.: Yang-Mills fields on quaternionic spaces. Nonlinearity **1**, 517–530 (1988)
15. Coble, A.B.: Algebraic Geometry and Theta Functions (Reprint of the 1929 Edition). A.M.S. Coll. Publ. vs 10 A.M.S, Providence (1982)
16. Cook, R.J., Thomas, A.D.: Line bundles and homogeneous matrices. Quart. J. Math. Oxford **30**, 423–429 (1979)
17. Costa, L., Miró-Roig, R.M.: Ulrich bundles on Veronese surfaces. In: Singularities, Algebraic Geometry, Commutative Algebra, and Related Topics, pp. 375–381 (2018)
18. Costa, L., Miró-Roig, R.M.: Instanton bundles vs Ulrich bundles on projective spaces. Beitr Algebra Geom. (2020). https://doi.org/10.1007/s13366-020-00506-7
19. Costa, L., Miró-Roig, R.M., Pons-Llopis, J.: Ulrich Bundles: From Commutative Algebra to Applications through Algebraic Geometry. Studies in Mathematics, vol. 77. De Gruyter (2021)
20. Corrigan, E., Goddard, P., Kent, A.: Some comments on the ADHM construction in 4k dimensions. Comm. Math. Phys. **100**, 1–13 (1985)
21. Cremona, L.: Sulle superficie gobbe di quatro grado. Mem. Accad. Scienze Ist. Bologna **8**, 235–250 (1868)
22. Leonard Eugene Dickson: Determination of all general homogeneous polynomials expressible as determinants with linear elements. Trans. Amer. Math. Soc. **22**, 167–179 (1921)
23. Dixon, A.: Note on the reduction of a ternary quartic to a symmetrical determinant. Proc. Cambridge Philos. Soc. **11**, 350–351 (1900–1902)
24. Eisenbud, D.: Homological algebra with an application to group representations. Trans. Amer. Math. Soc. **260**, 35–64 (1980)
25. Eisenbud, D., Schreyer, F.-O., (with an appendix by J. Weyman): resultants and Chow forms via exterior syzygies. J. Amer. Math. Soc. **16**, 537–579 (2003)
26. Ellia, P., Hirschowitz, A.: Génération de certain fibrés sur les spaces projectifs et application. J. Algebr. Geom. **1**, 531–547 (1992)
27. Fløystad, G., Kileel, J., Ottaviani, G.: The Chow form of the essential variety in computer vision. J. Symb. Comput. **86**, 97–119 (2018). https://doi.org/10.1016/j.jsc.2017.03.010

28. Grassmann, H.: Die stereometrischen Gleichungen dritten grades, und die dadurch erzeugten Oberflächen. J. Reine Angew. Math. **49**, 47–65 (1885)
29. Hartshorne, R.: Algebraic Geometry. Graduate Texts in Mathematics, vol. 52. Springer, New York (1977)
30. Hartshorne, R., Hirschowitz, A.: Cohomology of the general instanton bundle. Ann. scient. Éc. Norm. Sup. **15**, 365–390 (1982)
31. Hesse, O.: Ueber determinanten und ihre Anwendung in der Geometrie, insbesondere auf Curven vierter Ordnung. J. Reine Angew. Math. **49**, 243–264 (1855)
32. Hesse, O.: Ueber die Doppeltangenten der Curven vierter Ordnung. J. Reine Angew. Math. **49**, 279–332 (1855)
33. Kerner, D., Vinnikov, V.: Determinantal representations of singular hypersurfaces in ı^n. Adv. Math. **231**, 1619–1654 (2012)
34. Okonek, C., Schneider, M., Spindler, H.: Vector bundles on complex projective spaces. Progr. Math. **3** (1980)
35. Salamon, S.M.: Quaternionic structures and twistor spaces. In: Willmore, T.J., Hitchin, N.J. (eds.) Global Riemannian geometry (Durham, 1983). Ellis Horwood Ser. Math. Appl., pp. 65–74. Horwood, Chichester (1984)
36. Schur, F.: Über die durch collineare Grundgebilde erzeugten Curven und Flächen. Math. Annalen **18**, 1–32 (1881)
37. Ulrich, B.: Gorenstein rings and modules with high numbers of generators. Math. Z. **188**, 23–32 (1984)
38. van der Waerden, B.L.: Einführung in die Algebraische Geometrie. Grundlehren der Math. B XXXIII-XXIV (1930)

A Survey on Applications of a New Approach to Quillen Rational Homotopy Theory

Yves Félix[1], Mario Fuentes[2], and Aniceto Murillo[2(✉)]

[1] Institut de Mathématiques et Physique, Université Catholique de Louvain, Chemin du Cyclotron 2, 1348 Louvain-la-Neuve, Belgium
yves.felix@uclouvain.be

[2] Departamento de Álgebra, Geometría y Topología, Universidad de Málaga, 29080 Málaga, Spain
{m_fuentes,aniceto}@uma.es

Abstract. Recently, the classical Quillen approach to rational homotopy theory was extended to the non simply connected setting by means of a Quillen pair of adjoint functors

$$\mathbf{sset} \underset{\langle\cdot\rangle}{\overset{\mathfrak{L}}{\rightleftarrows}} \mathbf{cdgl}$$

between the categories of simplicial sets and complete differential graded Lie algebras. In this survey we present some applications, of this new approach. First we give an extension of the well know Baues-Lemaire conjecture, together with a short proof of this result. Then we describe how the new setting let us discover topological features hidden in the derivations of a given Lie algebra.

1 Introduction

The Quillen approach to rational homotopy theory [21] is based in a pair of functors,

$$\mathbf{sset}_1 \underset{\langle\cdot\rangle_Q}{\overset{\lambda}{\rightleftarrows}} \mathbf{dgl}_1 \tag{1}$$

between the categories of reduced simplicial sets, those with just one simplex in dimension 0 and 1, and that of positively graded differential graded Lie algebras. These functors induce equivalences when we restrict to the homotopy category of simply connected CW-complexes whose homotopy groups are rational

The authors has been partially supported by the grants PID2020-118753GB-I00 of the Spanish Government and PAIDI21-00827 of the Junta de Andalucía. *2010 Mathematics Subject Classification.* Primary: 55P62; Secondary: 55P30. *Key words and phrases*: Rational homotopy theory. Differential graded Lie algebras.

H. Herrero et al. (Eds.): RSME 2022, RSME 13, pp. 62–80, 2024.
https://doi.org/10.1007/978-3-031-62025-6_4

vector spaces (the so called rational CW-complexes) and the homotopy category of positively graded differential Lie algebras in which quasi-isomorphisms have been formally inverted [21, Theorem I].

Recently [5], the functors in (1) were extended "up to homotopy" by a pair of adjoint functors, model and realization, between the categories **sset** of simplicial sets and **cdgl** of complete differential graded Lie algebras (see next section for a brief compendium on the homotopy theory of this category). Moreover, these functors constitute a Quillen pair with respect to the model structure on **cdgl** inherited via transfer by the usual model structure on **sset**.

This can be used as a departure point to the extension of numerous and deep results describing the non torsion behaviour of the homotopy type of simply connected complexes to the general setting. The present notes surveys some examples of this kind.

It is convenient to remark at this point that in these notes we will often not distinguish simplicial sets from the topological spaces given by their realization which are therefore of the homotopy type of CW-complexes.

To begin with, we recall the well known *Baues-Lemaire Conjecture* [1] proved in the affirmative by M. Majewski [17] which connects the Sullivan and Quillen approach to rational homotopy theory as follows. Let X be a simply connected CW-complex of finite type, consider the positively graded dgl $\lambda(X)$ and replace it by $\mathbb{L}$, the *minimal Quillen model of* X. This is a free dgl of finite type, quasi-isomorphic to $\lambda(X)$ and whose differential is decomposable. We then may consider $\mathscr{C}^*(\mathbb{L})$, the commutative differential graded algebra (cdga henceforth) given by the usual cochain functor on $\mathbb{L}$. The Baues-Lemaire conjecture assert that this cdga is a Sullivan model of the complex X. That is, a free cdga whose differential also behaves in a recursive fashion and which characterizes the rational homotopy type of X. The new context given by the homotopy theory of cdgl's not only let us find a short proof of this result but also extend it to the non simply connected, nor finite type setting as we shall explicitly describe in Sect. 4.

Then we turn our attention to the description of the geometrical properties enclosed in the derivations of a given cdgl (see §3 for precise definitions and notation). Consider the dgl of derivations $\mathrm{Der}L$ of a given dgl, endowed with the usual Lie bracket and differential. When L is a Quillen model of the simply connected CW-complex X, then the "simply connected cover" of $\mathrm{Der}L$ is known to be a Lie model of the simply connected cover of $B\mathrm{aut}^*(X)$, the classifying space of the topological monoid of pointed self homotopy equivalences of X [25]. We present an extension of this result by showing how the classifying space of certain submonoids of free and pointed self homotopy equivalence of a given nilpotent CW-complex can be modeled in terms of connected lie algebras of derivations of the Lie model of X.

Finally, in Sect. 5 we describe how to extract further geometrical information from the *Maurer-Cartan set* (MC set henceforth) of $\mathrm{Der}L$ for some particular choices of L.

Explicitly, let π be a complete connected Lie algebra (with zero differential) and consider its *bigraded minimal Lie model* L (see §4 for precise definitions).

Then there exists a complete sub cdgl of $\mathrm{Der}(L)$ such that its MC set, module the gauge relation, is in one to one correspondence with the set of homotopy types of rational CW-complexes whose homotopy Lie algebra is isomorphic to π.

On the other hand, let H be a simply connected graded vector space and consider the free Lie algebra $L = \mathbb{L}(s^{-1}H)$ generated by the desuspension of H, with zero differential. Then, there is a natural action of the automorphisms of H on the MC set of a particular sub dgl of $\mathrm{Der}(L)$ such that the orbit set is in one to one correspondence with the set of homotopy types of rational simply connected CW-complexes sharing the same reduced homology H.

In a similar fashion, let A be (the augmentation ideal of) a simply connected cdga with zero differential and consider the dgl $L = \mathscr{L}_*(A)$ the "dual" of the cochain functor which is given by the classical Quillen functor on the coalgebra provided by the dual of A. Then, there is an action of the automorphisms of A on the MC set of a particular sub dgl of $\mathrm{Der}L$ such that the orbit set is in one to one correspondence with the set of homotopy types of simply connected CW-complexes sharing the same reduced rational cohomology algebra A.

This survey contains the main results of [8–10] and some key facts of [5]. Most proofs are omitted and others are simply outlined since, instead of being exhaustive, our main intention is to guide the reader in the most understandable way. Nevertheless, precise references are always given.

2 Homotopy Theory of Complete Differential Graded Lie Algebras

In this section we recall the basic facts from the homotopy theory of complete differential graded Lie algebras for which we refer to [5] for a detailed presentation, or to the original references [3, 4].

Throughout this paper we assume that $\mathbb{Q}$ is the base field and every considered vector space, possibly endowed with additional structures, are always graded over $\mathbb{Z}$. The *suspension* and *desuspension* of such a graded vector space V is denoted by sV and $s^{-1}V$ respectively. That is $(sV)_n = V_{n-1}$ and $(s^{-1}V)_n = V_{n+1}$ for any $n \in \mathbb{Z}$.

A *differential graded Lie algebra* (dgl) consists of a graded vector space $L = \oplus_{p\in\mathbb{Z}} L_p$ endowed with a bilinear product (Lie bracket),

$$[\,,\,]\colon L_p \otimes L_q \longrightarrow L_{p+q}, \quad p, q \in \mathbb{Z},$$

satisfying *antisymmetry* and *Jacobi identity*,

$$[x, y] = -(-1)^{|x||y|}[y, x],$$
$$(-1)^{|x||z|}[x, [y, z]] + (-1)^{|y||x|}[y, [z, x]] + (-1)^{|z||y|}[z, [x, y]] = 0,$$

and a differential d of degree -1 which is a graded derivation with respect to the Lie bracket and $d^2 = 0$. By $|x|$ we mean degree of x.

The commutator operator $[a,b] = a \otimes b - (-1)^{|a||b|} b \otimes a$ is a Lie bracket on $T(V)$, the tensor algebra on the graded vector space V. The *free Lie algebra* $\mathbb{L}(V)$ generated by V is the sub Lie algebra of $T(V)$ generated by V.

Given a differential graded Lie algebra L, a *Maurer-Cartan element* is an element $a \in L_{-1}$ satisfying the *Maurer-Cartan equation*

$$da + \frac{1}{2}[a,a] = 0.$$

The set of Maurer-Cartan, which we denote by $\mathrm{MC}(L)$, is clearly preserved by morphisms. Given $a \in \mathrm{MC}(L)$ consider the adjoint operator $\mathrm{ad}_a = [a, \cdot]$ and observe that the derivation $d_a = d + \mathrm{ad}_a$ is again a differential on L. The *component* of L at $a \in \mathrm{MC}(L)$, which we denote by L^a, is the connected sub dgl (i.e., concentrated in non negative degrees) of (L, d_a) given by

$$L_p^a = \begin{cases} L_p, & p \geq 1, \\ \ker d_a, & p = 0. \end{cases}$$

By a *filtration* of a dgl L we always mean a decreasing sequence of differential Lie ideals,

$$L = F^1 \supset \cdots \supset F^n \supset F^{n+1} \supset \ldots$$

such that $[F^p, F^q] \subset F^{p+q}$ for $p, q \geq 1$. For instance, the lower central series of L,

$$L^1 \supset \cdots \supset L^n \supset L^{n+1} \supset \ldots,$$

where $L^1 = L$ and $L^n = [L, L^{n-1}]$ for $n > 1$, is a filtration for any dgl.

A *complete differential graded Lie algebra* (cdgl) is a dgl L equipped with a filtration $\{F^n\}_{n\geq 1}$ for which the natural map

$$L \xrightarrow{\cong} \varprojlim_n L/F^n$$

is an isomorphism. Note that, if L is simply connected or more generally, nilpotent, then L is complete with respect to the filtration given by the lower central series. A cdgl *morphism* is a dgl morphism which preserves the filtrations. We denote by **cdgl** the corresponding category. A *complete graded Lie algebra* (cgl) is a cdgl endowed with the zero differential.

If L is a dgl filtered by $\{F^n\}_{n\geq 1}$, its *completion* is the dgl

$$\widehat{L} = \varprojlim_n L/F^n$$

which is always complete with respect to the filtration

$$\widehat{F^n} = \ker(\widehat{L} \to L/F^n).$$

If no specific filtration is given, the completion of a generic dgl is always taken over the lower central series. In particular, if $\mathbb{L}(V)$ denotes the free Lie algebra generated by the graded vector space V, the completion of a dgl of the form $(\mathbb{L}(V), d)$ is the cdgl

$$\widehat{\mathbb{L}}(V) = \varprojlim_n \mathbb{L}(V)/\mathbb{L}(V)^n.$$

The Baker-Campbell-Hausdorff (BCH) product $*$ equips any cgl, in particular the degree zero elements L_0 of any given cdgl L, with a group structure. Moreover, there is an action of $(L_0, *)$ on $\mathrm{MC}(L)$ called the *gauge action* which is defined by

$$x \mathcal{G} a = e^{\mathrm{ad}_x}(a) - \frac{e^{\mathrm{ad}_x} - \mathrm{id}}{\mathrm{ad}_x}(dx) = \sum_{n\geq 0} \frac{\mathrm{ad}_x^n(a)}{n!} - \sum_{n\geq 0} \frac{\mathrm{ad}_x^n(dx)}{(n+1)!}, \quad x \in L_0, \quad a \in \mathrm{MC}(L).$$

We denote by $\widetilde{\mathrm{MC}}(L) = \mathrm{MC}(L)/\mathcal{G}$ the orbit set.

The homotopy theory of cdgl's lies in the existence of a pair of adjoint functors [5, Chapter 7], *model* and *realization*,

$$\mathbf{sset} \underset{\langle\cdot\rangle}{\overset{\mathfrak{L}}{\rightleftarrows}} \mathbf{cdgl}\,. \tag{2}$$

which is based on the cosimplicial cdgl $\mathfrak{L}_\bullet$: for each $n \geq 1$, there is a free cdgl $\mathfrak{L}_n = (\widehat{\mathbb{L}}(s^{-1}\Delta^n), d)$ in which $s^{-1}\Delta^n$ together with the linear part of the differential d is the (desuspension) of the rational simplicial chain complex of the standard n-simplex Δ^n, and the vertices correspond to Maurer-Cartan elements. Moreover, the usual cofaces and degeneracies on the cosimplicial chain complex $s^{-1}\Delta^\bullet$ extend to cofaces and codegeneracies on $\mathfrak{L}_\bullet$.

Then, the realization of a given cdgl is defined as

$$\langle L \rangle = \mathrm{Hom}_{\mathbf{cdgl}}(\mathfrak{L}_\bullet, L).$$

On the other hand, if we denote $\mathfrak{L}_{\Delta^n} = \mathfrak{L}_n$, the construction of $\mathfrak{L}_\bullet$ can be extended to any simplicial set X by defining its dgl model as

$$\mathfrak{L}_X = \varinjlim_{\sigma\in X} \mathfrak{L}_{\Delta^{|\sigma|}}.$$

The set of 0-simplices and path components of $\langle L \rangle$ coincide, respectively with $\mathrm{MC}(L)$ and $\widetilde{\mathrm{MC}}(L)$. Moreover, if $\langle L \rangle^a$ denotes the path component of $\langle L \rangle$ containing the MC element a, we have:

$$\langle L \rangle^a \simeq \langle L^a \rangle, \quad \langle L \rangle \simeq \amalg_{a\in\widetilde{\mathrm{MC}}(L)} \langle L^a \rangle. \tag{3}$$

If L is connected, and for any $n \geq 1$, we have group isomorphisms

$$\pi_n\langle L \rangle \cong H_{n-1}(L)$$

where the group structure in $H_0(L)$ is considered with the BCH product.

The realization functor is also closely related to the so called *Deligne-Getzler-Hinich groupoid* functor [14,16]

$$\mathrm{MC}_\bullet\colon \mathbf{cdgl} \longrightarrow \mathbf{sset}$$

which assigns to each cdgl L the simplicial set $\mathrm{MC}_\bullet(L)$ given by the MC set of the simplicial cdgl $\mathscr{A}_\bullet\widehat{\otimes}L$, in which $\mathscr{A}_\bullet = \mathscr{A}(\Delta^\bullet)$ denotes the simplicial commutative differential graded algebra of PL-differential forms on the standard simplices. Recall that given any cdga A and any cdgl L the complete tensor product $A\widehat{\otimes}L = \varprojlim_n A\otimes L/F^n$ has a natural cdgl structure. It turns out [6, Theorem 0.2] that $\langle L\rangle$ is a deformation retract of $\mathrm{MC}_\bullet(L)$.

To embed the category of cdgl's in a suitable homotopy theoretical framework we transfer the usual model structure on **sset** via (2) to endow **cdgl** with a cofibrantly generated model category structure in which: a cdgl morphism $f\colon A\to B$ is a fibration if it is surjective in non negative degrees; f is a weak equivalence if $\widetilde{\mathrm{MC}}(f)\colon \widetilde{\mathrm{MC}}(A)\xrightarrow{\cong}\widetilde{\mathrm{MC}}(B)$ is a bijection and $f^a\colon A^a\xrightarrow{\simeq} B^{f(a)}$ is a quasi-isomorphism for every $a\in\widetilde{\mathrm{MC}}(A)$; finally f is a cofibration if it has the left lifting property with respect to trivial fibrations. As an immediate consequence we deduce that the model and realization functors form a Quillen pair. In particular, they induce adjoint functors in the homotopy categories,

$$\mathrm{Ho}\,\mathbf{sset} \underset{\langle\cdot\rangle}{\overset{\mathfrak{L}}{\rightleftarrows}} \mathrm{Ho}\,\mathbf{dgl},$$

and both preserve weak equivalences and homotopies. It turns out [10, Corollary 2.3] that these functors extend the original Quillen equivalences

$$\mathrm{Ho}\,\mathbf{sset}_1^{\mathbb{Q}} \underset{\langle\cdot\rangle_Q}{\overset{\lambda}{\rightleftarrows}} \mathrm{Ho}\,\mathbf{dgl}_1.$$

Here, $\mathrm{Ho}\,\mathbf{sset}_1^{\mathbb{Q}}$ denotes the homotopy category of rational reduced simplicial sets.

In particular, if L is a connected cdgl, $\langle L\rangle$ is homotopy equivalent to $\langle L\rangle_Q$. Conversely if X is a simply connected simplicial set and a is any of its vertices, $\mathfrak{L}_X^a$ is quasi-isomorphic to $\lambda(X)$ where λ is the classical Quillen dgl model functor [21]. Moreover, see [5, Theorem 11.14], for any connected simplicial set X of finite type, $\langle\mathfrak{L}_X^a\rangle$ is weakly homotopy equivalent to $\mathbb{Q}_\infty X$ the Bousfield-Kan $\mathbb{Q}$-completion of X [2]. Recall that, whenever X is nilpotent, $\mathbb{Q}_\infty X$ and has the homotopy type of $X_{\mathbb{Q}}$, the rationalization of X.

In this context, a *model* of a connected cdgl L is a connected cdgl of the form $(\widehat{\mathbb{L}}(V),d)$ together with a quasi-isomorphism (and hence a weak equivalence)

$$(\widehat{\mathbb{L}}(V),d)\xrightarrow{\simeq} L.$$

If d is decomposable we say that $(\widehat{\mathbb{L}}(V),d)$ is the *minimal model* of L and is unique up to cdgl isomorphism.

Given X be a connected simplicial set and a any of its vertices, the *minimal model of* X is the minimal model of $\mathfrak{L}_X^a$.

If $(\widehat{\mathbb{L}}(V), d)$ is the minimal model of X then, see [5, Proposition 8.35], $sV \cong \widetilde{H}_*(X;\mathbb{Q})$ and, provided X of finite type, $sH_*(\widehat{\mathbb{L}}(V), d) \cong \pi_*(\mathbb{Q}_\infty X)$. Again, the group $H_0(\widehat{\mathbb{L}}(V), d)$ is considered with the BCH product. Note that if X is simply connected, the minimal model of X is isomorphic to its classical Quillen minimal model.

We finish this brief summary by relating our realization functor with the so called the *Deligne-Getzler-Hinich groupoid* functor [14,16] $\mathrm{MC}_\bullet\colon \mathbf{cdgl} \to \mathbf{sset}$. Given L a cdgl, $\mathrm{MC}_\bullet(L) = \mathrm{MC}(\mathscr{A}_\bullet \widehat{\otimes} L)$ is the simplicial set of Maurer-Cartan elements of the simplicial cdgl

$$\mathscr{A}_\bullet \widehat{\otimes} L = \varprojlim_n \mathscr{A}_\bullet \otimes L/F^n$$

in which $\mathscr{A}_\bullet$ denotes the simplicial commutative differential graded algebra of PL-differential forms on the standard simplices. It turns out [6, Theorem 0.2] that $\langle L \rangle$ is a defomration retract of $\mathrm{MC}_\bullet(L)$.

3 The Baues-Lemaire Conjecture

The commutative approach to rational homotopy theory is due to D. Sullivan, see [24] or the standard reference [11], and lies in the construction of a pair of adjoint functors,

$$\mathbf{sset} \underset{\langle \cdot \rangle_S}{\overset{\mathscr{A}}{\rightleftarrows}} \mathbf{cdga}. \tag{4}$$

between the category of simplicial sets and that of cdga's, analogue to (2), which is also based on the existence of the "universal" simplicial cdga $\mathscr{A}_\bullet = \mathscr{A}(\Delta^\bullet)$. Explicitly, given a cdga A, its Sullivan realization is the simplicial set

$$\langle A \rangle_S = \mathrm{Hom}_{\mathbf{cdga}}(A, \mathscr{A}_\bullet).$$

On the other hand, given X any simplicial set, $\mathscr{A}(X)$ is the cdga of PL-forms on X.

Given any connected simplicial set X, the cdga $\mathscr{A}(X)$ can be replaced by a simpler, homotopy equivalent cdga, called a *Sullivan model of* X. This is a free cdga $(\Lambda V, d)$ generated by a graded vector space positively graded which is equipped with a (homogeneous) ordered basis in which the differential d "acts recursively": on each element v of such basis dv is a polynomial on ΛV involving only elements less than v.

The pair (4) induces equivalences between the homotopy category of rational, nilpotent CW-complexes of finite type (over $\mathbb{Q}$) and the homotopy category of connected cdga's (non negatively graded and having $\mathbb{Q}$ as elements of degree 0) admitting a Sullivan model of finite type.

Classically, the connection between the commutative and Lie approach is given by the cochain functor

$$\mathscr{C}^*\colon \mathbf{dgl}_0^f \longrightarrow \mathbf{cdga}_0^f$$

which runs from the category of connected dgl's (non negatively graded) of finite type to that of connected cdga's of finite type. It is defined by

$$\mathscr{C}^*(L) \cong (\land (sL)^\sharp, d)$$

in which $\sharp$ denotes the dual and $d = d_1 + d_2$ where

$$\langle d_1 v, sx\rangle = (-1)^{|v|}\langle v, sdx\rangle,$$
$$\langle d_2 v, sx \land sy\rangle = (-1)^{|y|+1}\langle v, s[x,y]\rangle,$$

with $v \in (sL)^\sharp$ and $x, y \in L$.

One then may ask whether this connection also works at the topological level: let X be a simply connected CW-complex of finite type. Consider the simply connected or reduced dgl $\lambda(X)$ and replace it by a *Quillen model of* X. This is a reduced, free dgl $L = (\mathbb{L}, d)$ of finite type related to $\lambda(X)$ by a quasi-isomorphism $L \xrightarrow{\simeq} \lambda(X)$. Is it $\mathscr{C}(L)$ a Sullivan model of X? This was conjectured to be true in [1, Conjecture 3.5] and it is known as the *Baues-Lemaire Conjecture*. It was finally proved later on by M. Majewski in [17] by a long, deep and technical procedure. However, with the new framework given by the homotopy theory of cdgl's we can easily reprove it:

Theorem 3.1. *The Baues-Lemaire conjecture holds.*

Sketch of proof. By the equivalences induced by the functors in (4) the statement is equivalent to show that $\langle \mathscr{C}^*(L)\rangle_S$ has the homotopy type of $X_\mathbb{Q}$. A direct proof can be found in [4, Theorem 8.1]. Here, we sketch a different procedure: since $\lambda(X)$ and $\mathfrak{L}^a(X)$ are quasi-isomorphic there is no difference between a Quillen model or a Lie model of X. Now, let A be any connected cdga bounded above ($A^{>n} = 0$ for some n). Then, if we denote by **cga** the category of commutative graded algebras with no differential, it is obvious that

$$\operatorname{Hom}_{\mathbf{cga}}(\mathscr{C}^*(L), A) = \operatorname{Hom}\big((sL)^\sharp, A\big) = (A \otimes L)_{-1}$$

where the unadorned Hom stands for linear maps of degree 0. Now, it is straightforward to check the classical fact by which the image of the inclusion

$$\operatorname{Hom}_{\mathbf{cdga}}(\mathscr{C}^*(L), A) \hookrightarrow \operatorname{Hom}_{\mathbf{cga}}(\mathscr{C}^*(L), A) \cong (A \otimes L)_{-1}$$

is precisely the MC set of $A \otimes L$. Note also that, since L is simply connected, $A \otimes L = A \widehat{\otimes} L$. Therefore,

$$\langle \mathscr{C}^*(L)\rangle_S = \operatorname{Hom}_{\mathbf{cdga}}(\mathscr{C}^*(L), \mathscr{A}_\bullet) = \operatorname{MC}(\mathscr{A}_\bullet \widehat{\otimes} L) = \operatorname{MC}_\bullet(L).$$

Finally recall that $\langle L\rangle$, which is of the homotopy type of $X_\mathbb{Q}$, is a deformation retract of $\operatorname{MC}_\bullet(L)$ $\square$

We finish this section by showing how the new context let us extend this conjecture to the non simply connected case. Let L be the minimal Lie model of a connected simplicial set of finite type. While L may not be of finite type, L/L^n always is for any $n \geq 1$. Then:

Theorem 3.2. *[5, Theorem 10.8]* $\varinjlim_n \mathscr{C}^*(L/L^n)$ *is a Sullivan model of* X.

Observe that if L is simply connected the lower central series of L is always finite (degreewise) and thus $\varinjlim_n \mathscr{C}^*(L/L^n) = \mathscr{C}^*(L)$ which again recovers the Baues-Lemaire conjecture.

4 Derivations of Cdgl's and Certain Monoids of Self Homotopy Equivalences

Given a dgl L we denote by $\mathrm{Der}L$ the dgl consisting of its *derivations*. For each $n \in \mathbb{Z}$, $\mathrm{Der}_n L$ are linear maps $\theta\colon L \to L$ of degree n such that

$$\theta[a,b] = [\theta(a), b] + (-1)^{|a|n}[a, \theta(b)], \quad \text{for } a, b \in L.$$

The Lie bracket and differential are given by,

$$[\theta, \eta] = \theta \circ \eta - (-1)^{|\theta||\eta|}\eta \circ \theta, \quad D\theta = [d, \mathrm{id}_L] = d \circ \theta - (-1)^{|\theta|}\theta \circ d.$$

From the topological point of view, and in the simply connected context, this object is closely related to the understanding of the rational homotopy type of certain classifying spaces. Indeed, let X be a pointed CW-complex and consider the free and pointed mapping spaces $\mathrm{map}(X,X)$ and $\mathrm{map}^*(X,X)$. The *evaluation fibration*,

$$\mathrm{map}^*(X,X) \longrightarrow \mathrm{map}(X,X) \longrightarrow X,$$

which consists on evaluating each continuous map on the base point, restricts to another fibration sequence

$$\mathrm{aut}^*(X) \longrightarrow \mathrm{aut}(X) \longrightarrow X \tag{5}$$

where $\mathrm{aut}(X)$ (respec. $\mathrm{aut}^*(X)$) denotes the topological monoid of self homotopy equivalences (respec. pointed self homotopy equivalences) of X.

This fibration can be extended on the right to another fibration sequence

$$X \longrightarrow B\mathrm{aut}^*(X) \longrightarrow B\mathrm{aut}(X) \tag{6}$$

where $B\mathrm{aut}(X)$ and $B\mathrm{aut}^*(X)$ denote the classifying spaces of $\mathrm{aut}(X)$ and $\mathrm{aut}^*(X)$ respectively. It is well known [19,23] that (6) is the *universal classifying fibration* with respect to fibrations sequences of fibre X. That is, for any CW-complex Y there is a bijection

$$[Y, B\mathrm{aut}(X)] \cong \mathcal{F}\mathrm{ib}(Y, X)$$

where $\mathcal{F}\mathrm{ib}(Y,X)$ denotes the set of homotopy classes of fibration sequences of base Y and fibre X. This correspondence associates to each homotopy class $[f]$ in $[Y, B\mathrm{aut}(X)]$ the pullback of the universal fibration over f.

Let now X be simply connected and observe that the universal cover of the universal fibration can be identified to the fibration sequence

$$X \longrightarrow B\mathrm{aut}_1^*(X) \longrightarrow B\mathrm{aut}_1(X)$$

obtained, as above, by extending to the right the restriction of (5)

$$\mathrm{aut}_1^*(X) \longrightarrow \mathrm{aut}_1(X) \longrightarrow X \tag{7}$$

to the component $\mathrm{aut}_1(X)$ of $\mathrm{aut}(X)$ consisting of the self homotopy equivalences of X homotopic to the identity.

On the algebraic side, let L be the classical Quillen model of X and consider the dgl sequence

$$L \xrightarrow{\mathrm{ad}} \widetilde{\mathrm{Der}L} \xrightarrow{j} \widetilde{\mathrm{Der}L}\tilde{\times} sL \tag{8}$$

defined as follows:

- $\widetilde{\mathrm{Der}L}$ is the "simply connected cover of $\mathrm{Der}L$. This is the sub dgl of $\mathrm{Der}L$ consisting of all derivations in degree greater than 1 and the cycles in degree 1.
- ad is the adjoint operator sending each element $x \in L$ to the derivation ad_x.
- $\widetilde{\mathrm{Der}L}\tilde{\times} sL$ is the dgl structure on the product $\widetilde{\mathrm{Der}L} \times sL$ where $\widetilde{\mathrm{Der}L}$ is a sub dgl, sL is an abelian Lie algebra and

$$Dsx = -sdx + \mathrm{ad}_x, \quad [\theta, sx] = (-1)^{|\theta|} s\theta(x), \quad x \in L, \quad \theta \in \mathrm{Der}L.$$

- j is just the inclusion.

Then:

Theorem 4.1. [25, §7] *The dgl sequence*

$$L \xrightarrow{\mathrm{ad}} \widetilde{\mathrm{Der}L} \xrightarrow{j} \widetilde{\mathrm{Der}L}\tilde{\times} sL$$

is a Quillen model of the fibration

$$X \longrightarrow B\mathrm{aut}_1^*(X) \longrightarrow B\mathrm{aut}_1(X)$$

Within the new approach to Quillen rational homotopy theory, this result can be extended, although not in complete generality. Indeed, and even if X is simply connected, $B\mathrm{aut}(X)$ might not admit any Lie model due to the possible complexity of its fundamental group $\pi_1 B\mathrm{aut}(X) = \pi_0\mathrm{aut}(X)$ which is the group of homotopy classes of self homotopy equivalences of X, denoted by $\mathcal{E}(X)$ henceforth.

Example 4.2. Let $S^n_{\mathbb{Q}}$ be the rationalization of the n-dimensional sphere. Then the classifying space $B\mathrm{aut}(S^n_{\mathbb{Q}})$ is not in the image of the realization functor $\langle\cdot\rangle$.

Otherwise we would have $B\mathrm{aut}(S^n_{\mathbb{Q}}) \simeq \langle L\rangle$ for a given connected cdgl L and a group isomorphism between $\pi_1 B\mathrm{aut}(S^n_{\mathbb{Q}}) = \mathcal{E}(S^n_{\mathbb{Q}})$ with $H_0(L)$ endowed with the BCH product.

On the one hand it is easy to see that $\mathcal{E}(S^n_{\mathbb{Q}})$ is the group of automorphisms of a one dimensional vector space which is identified with the multiplicative group $\mathbb{Q}^* = \mathbb{Q} - \{0\}$.

On the other hand, if $\psi\colon \mathbb{Q}^* \xrightarrow{\cong} H_0(L)$ is an isomorphism choose $a = \psi(2)$ and recall that, for the BCH product, $\mu a * \nu a = (\lambda+\mu)a$ for $\mu,\nu \in \mathbb{Q}$. Thus, given $\lambda \in \mathbb{Q}^*$ with $\psi(\lambda) = \frac{1}{2}a$ we would have $\psi(\lambda^2) = \frac{1}{2}a * \frac{1}{2}a = a = \psi(2)$. Then $\lambda^2 = 2$ which is a contradiction.

Nevertheless we may proceed as follows: given X any CW-complex and $G \subset \mathcal{E}(X)$ any subgroup, we consider the subgroups $\mathrm{aut}_G(X), \mathrm{aut}^*_G(X) \subset \mathrm{aut}(X)$ defined by

$$\mathrm{aut}_G(X) = \{f \in \mathrm{aut}(X),\ [f] \in G\},\quad \mathrm{aut}^*_G(X) = \{f \in \mathrm{aut}^*(X),\ [f] \in G\}.$$

By restricting once again the fibrations (5) and (6) we get a long fibration sequence

$$\mathrm{aut}^*_G(X) \longrightarrow \mathrm{aut}_G(X) \longrightarrow X \longrightarrow B\mathrm{aut}^*_G(X) \longrightarrow B\mathrm{aut}_G(X)$$

whose last part is also universal in the following sense: given a CW-complex Y denote by $\mathcal{F}\mathrm{ib}_G(Y,X)$ the homotopy classes of fibration sequences over Y, with fibre X and such that the image of the natural morphism $\pi_1(Y) \to \mathcal{E}(X)$ lies in G. Then [13], there is a bijection

$$[Y, B\mathrm{aut}_G(X)] \cong \mathcal{F}\mathrm{ib}_G(Y,X)$$

also given by taking the pullback of the fibration $X \to B\mathrm{aut}^*_G(X) \to B\mathrm{aut}_G(X)$ over a given homotopy class in $[Y, B\mathrm{aut}_G(X)]$.

Assume now that X is a finite nilpotent CW-complex and that G acts nilpotently on $H_*(X)$. A deep result of Dror and Zabrodsky [7] asserts that the spaces $B\mathrm{aut}_G(X)$ and $B\mathrm{aut}^*_G(X)$ are nilpotent. In particular G is also a nilpotent group and we denote by $G_{\mathbb{Q}} \subset \mathcal{E}(X_{\mathbb{Q}})$ its classical rationalization.

On the other hand, consider $\mathcal{E}^*(X_{\mathbb{Q}})$ the group of pointed homotopy classes of pointed homotopy equivalences of $X_{\mathbb{Q}}$. As for any other space, there is a natural action of $\pi_1(X_{\mathbb{Q}})$ on $\mathcal{E}^*(X_{\mathbb{Q}})$, which induces a bijection

$$\mathcal{E}^*(X_{\mathbb{Q}})/\pi_1(X_{\mathbb{Q}}) \cong \mathcal{E}(X_{\mathbb{Q}}),$$

and is modeled as follows, see [8, §2]: if we denote by L the minimal Lie model of X, the group $\mathcal{E}^*(L) = \mathrm{aut}(L)/\sim$ of homotopy classes of automorphisms of L is naturally identified with the group $\mathcal{E}^*(X_{\mathbb{Q}})$. Then, on the algebraic side, the action of $\pi_1(X_{\mathbb{Q}})$ on $\mathcal{E}^*(X_{\mathbb{Q}})$ is identified with the action

$$H_0(L) \times \mathcal{E}^*(L) \longrightarrow \mathcal{E}^*(L),\qquad ([x],[\varphi]) \mapsto [e^{\mathrm{ad}_x} \circ \varphi],$$

where $e^{\mathrm{ad}_x} = \sum_{n\geq 0} \frac{\mathrm{ad}_x^n}{n!}$. Therefore, there is a group isomorphism

$$\mathcal{E}^*(L)/H_0(L) \cong \mathcal{E}(X_{\mathbb{Q}}).$$

We then consider the subgroup

$$\mathcal{G} \subset \mathcal{E}^*(L),$$

of homotopy classes of automorphisms of L such that

$$\mathcal{G}/H_0(L) \cong G_{\mathbb{Q}}$$

under the above isomorphism. We also denote

$$\mathrm{aut}_{\mathcal{G}}(L) = \{\varphi \in \mathrm{aut}(L),\ [\varphi] \in \mathcal{G}\},$$

which is clearly a subgroup of $\mathrm{aut}(L)$.

At this point it is convenient to recall the following: given a group K we denote its commutator by curved brackets:

$$(x, y) = xyx^{-1}y^{-1}$$

and consider its lower central series,

$$K = K^1 \supset K^2 \supset \cdots \supset K^i \supset K^{i+1} \supset \dots$$

where $K^i = (K^{i-1}, K)$ for $i \geq 2$. The group K is *pronilpotent* if the natural map $K \xrightarrow{\cong} \varprojlim_n K/K^n$ is an isomorphism. On the other hand recall that K is said to be 0-local, uniquely divisible or rational if for any natural $n \geq 1$ the map $K \to K$, $g \mapsto g^n$, is a bijection.

Definition 4.3. A group K is *Malcev $\mathbb{Q}$-complete* (or complete, for short) if it is pronilpotent and for each $n \geq 1$ the abelian group K^n/K^{n+1} is a $\mathbb{Q}$-vector space.

We then can prove:

Proposition 4.4. [8, Lemma 8.1] $\mathrm{aut}_{\mathcal{G}}(L)$ *is a complete group.*

We also recall that the original result of Malcev [18] identifying rational, finitely generated nilpotent groups with finitely generated nilpotent Lie algebras can be extended to obtain an isomorphism between the categories of complete groups and complete Lie algebras (cgl henceforth, ungraded and with no differentials), see for instance [12, §8.2.8].

In particular, the complete Lie algebra $\mathrm{aut}_{\mathcal{G}}(L)$ is identified with a certain sub cdgl of $\mathrm{Der}_0 L$ via the logarithm and the exponential map. Indeed, given $\varphi \in \mathrm{aut}_{\mathcal{G}}(L)$, consider the derivation of L of degree 0 given by

$$\log(\varphi)(x) = \sum_{n\geq 1}(-1)^{n+1}\frac{(\varphi - \mathrm{id}_L)^n(x)}{n}, \qquad x \in L.$$

Define $\mathrm{Der}_0^{\mathcal{G}} L$ as Im log so that there are bijections

$$\mathrm{aut}_{\mathcal{G}}(L) \underset{\exp}{\overset{\log}{\rightleftarrows}} \mathrm{Der}_0^{\mathcal{G}} L \tag{9}$$

in which

$$\exp(\theta) = e^{\theta} = \sum_{n\geq 0} \frac{\theta^n}{n!}, \quad \theta \in \mathrm{Der}_0^{\mathcal{G}} L.$$

We add to $\mathrm{Der}_0^{\mathcal{G}} L$ all derivations in positive degrees to get:

Definition 4.5. Let $\mathrm{Der}^{\mathcal{G}} L$ be the connected sub cdgl of $\mathrm{Der} L$ defined by:

$$\mathrm{Der}_k^{\mathcal{G}} L = \begin{cases} \mathrm{Der}_k L & \text{if } k \geq 1, \\ \mathrm{Der}_0^{\mathcal{G}} L & \text{if } k = 0, \end{cases}$$

Then, the extension of Theorem 4.1 is the following:

Theorem 4.6. *[8, Theorem 7.13] the cdgl sequence, defined as in (8),*

$$L \xrightarrow{\mathrm{ad}} \mathrm{Der}^{\mathcal{G}} L \xrightarrow{j} \mathrm{Der}^{\mathcal{G}} L \widetilde{\times} sL$$

is a Lie model of the universal fibration

$$X \longrightarrow B\mathrm{aut}_G^*(X) \longrightarrow B\mathrm{aut}_G(X).$$

Among others, immediate consequences of this result is the explicit description of the rationalization of the group G and its classifying space BG also in terms of derivations of L.

Corollary 4.7. *(a)* $G_{\mathbb{Q}} \cong H_0(\mathrm{Der}^{\mathcal{G}} L)/\mathrm{Im}\, H_0(\mathrm{ad})$.
(b) A Lie model of BG is given by $\mathrm{Der}_0^{\mathcal{G}} L \oplus R$ where R denotes a complement of the 1-cycles of $\mathrm{Der}^{\mathcal{G}} L \widetilde{\times} sL$.

Proof. Consider the fibration sequence

$$\mathrm{aut}_1(X) \longrightarrow \mathrm{aut}_G(X) \longrightarrow G$$

given by $f \in \mathrm{aut}_G(X) \mapsto [f] \in G$, considering G as a discrete space. This gives rise to another fibration

$$B\mathrm{aut}_1(X) \longrightarrow B\mathrm{aut}_G(X) \longrightarrow BG \tag{10}$$

which can be thought also as fibring $B\mathrm{aut}_G(X)$ over its first Postnikov piece. However Postinkov pieces are easy to model: given M a connected cdgl and $n \geq 1$, consider the short exact sequence

$$M_{>n} \oplus Z_n \longrightarrow M \longrightarrow M/(M_{>n} \oplus Z_n)$$

where $Z_n \subset M_n$ is the subspace of cycles. Then [5, Proposition 12.43], the realization of this sequence is of the homotopy type of the fibration of $\langle M \rangle$ over its nth Postnikov stage. In particular, in view of Theorem 4.6, the fibration (10) is modeled by

$$\widetilde{\mathrm{Der} L\tilde{\times} sL} \hookrightarrow \mathrm{Der}^{\mathcal{G}} L\tilde{\times} sL \to (\mathrm{Der}^{\mathcal{G}} L\tilde{\times} sL)/(\widetilde{\mathrm{Der} L\tilde{\times} sL})$$

where $\widetilde{\mathrm{Der} L\tilde{\times} sL}$ denotes the simply connected cover of $\mathrm{Der} L\tilde{\times} sL$. A short computation let us observe that

$$(\mathrm{Der}^{\mathcal{G}} L\tilde{\times} sL)/(\widetilde{\mathrm{Der} L\tilde{\times} sL}) = \mathrm{Der}_0^{\mathcal{G}} L \oplus R$$

so that (b) follows. Finally $G_{\mathbb{Q}} = \pi_1(BG_{\mathbb{Q}})$ which, by (b) is precisely $H_0(\mathrm{Der}_0^{\mathcal{G}} L \oplus R)$. It is a straightforward computation to check that this coincides with $H_0(\mathrm{Der}^{\mathcal{G}} L)/\mathrm{Im}\, H_0(\mathrm{ad})$.

Example 4.8. We finish this section illustrating all of the above with a particular instance. Let $X = S^n \vee S^n \vee S^n$ be the wedge of three n-dimensional spheres with $n \geq 1$. The minimal model of X is $L = (\mathbb{L}(x, y, z), 0)$ in which $|x| = |y| = |z| = n-1$. In this case, as the differential is trivial, $\mathcal{E}(X_{\mathbb{Q}}) \cong \mathrm{aut}(L)$. However, for degree reasons, each automorphism of L is determined by an automorphism of the generating vector space $\mathrm{Span}(x, y, z)$ so that

$$\mathcal{E}(X_{\mathbb{Q}}) \cong \mathrm{GL}(3; \mathbb{Q}).$$

Choose the (already rational) subgroup $G \subset \mathcal{E}(X_{\mathbb{Q}})$ which, by the above identification is generated by the automorphism $\varphi \in \mathrm{aut}(L)$ given by the matrix

$$\begin{bmatrix} 1 & 1 & 1 \\ 0 & 1 & 1 \\ 0 & 0 & 1 \end{bmatrix}.$$

That is,

$$\varphi(x) = x + y + z, \quad \varphi(y) = y + z, \quad \varphi(z) = z.$$

Clearly $G = G_{\mathbb{Q}}$ acts nilpotently on $H_*(X; \mathbb{Q}) = H_*(X_{\mathbb{Q}}) \cong \mathbb{Q} \oplus s^{-1}\mathrm{Span}(x, y, z)$ and it is a nilpotent group. In this particular example $\mathcal{G} \cong G$ and $\mathrm{Der}_0^{\mathcal{G}} L$ is then generated by the derivation of L given by

$$\theta = \log(\varphi) = \begin{bmatrix} 0 & 1 & \frac{1}{2} \\ 0 & 0 & 1 \\ 0 & 0 & 0 \end{bmatrix}.$$

That is,

$$\theta(x) = y + \frac{z}{2}, \quad \theta(y) = z, \quad \theta(z) = 0.$$

All the results above can then be applied to L and $\mathrm{Der}^{\mathcal{G}} L = \mathrm{Der}_{\geq 1} L \oplus \mathrm{Span}(\theta)$. In particular, also for degree reasons BG is modeled by the Lie algebra generated by θ and is therefore homotopy equivalent to the circle.

5 MC Sets of Cdgl's of Derivations and Moduli Spaces of Certain Rational Homotopy Types

In the past section we've seen how certain connected covers of cdgl's of derivations model particular monoids of self homotopy equivalences. Here, we do not truncate theses cdgl's and study, from a geometrical point of view, their MC sets.

First of all it is convenient to remark that only complete dgl's are susceptible of being realized and in some cases, even if L is simply connected, $\mathrm{Der}L$ may fail to be complete. An easy example if the following: let $L = (\mathbb{L}(x,y),0)$ with $|x| = |y| = 2$, and consider $\theta_1, \theta_2, \theta_3 \in \mathrm{Der}_0 L$ defined by

$$\theta_1(x) = x,\ \theta_1(y) = -y, \qquad \theta_2(x) = y,\ \theta_2(y) = 0, \qquad \theta_3(x) = 0,\ \theta_3(y) = x,$$

and then extending these map to L as derivations. It is a simple computation to show that

$$[\theta_1, \theta_2] = -2\theta_2, \quad [\theta_1, \theta_3] = 2\theta_3, \quad [\theta_2, \theta_3] = -\theta_1.$$

Hence, $\theta_i \in L^n$ for any n and any i. That is, these derivations live in the kernel of the natural map $\mathrm{Der}L \to \varprojlim_{n\geq 1} \mathrm{Der}L/F^n$ and thus it is not an isomorphism. That is, $\mathrm{Der}L$ is not complete.

We then consider any sub cdgl M of $(\mathrm{Der}L, D)$ and observe the following facts:

First, by definition, $\delta \in M_{-1}$ is a Maurer-Cartan element if

$$D(\delta) = d \circ \delta + \delta \circ d = -\frac{1}{2}[\delta, \delta] = -\delta^2.$$

In other terms,

$$\mathrm{MC}(M) = \{\delta \in M_{-1} \text{ such that } d + \delta \text{ is a differential in } L\}. \tag{11}$$

We also have the following characterization of the gauge relation:

Proposition 5.1. *Two Maurer-Cartan elements $\delta, \eta \in \mathrm{MC}(M)$ are gauge related if and only if there exists an isomorphism of the form*

$$e^\theta \colon (L, d+\delta) \xrightarrow{\cong} (L, d+\eta)$$

with $\theta \in M_0$. Furthermore the gauge action is given by $\theta \,\mathcal{G}\, \delta = \eta$.

Proof. Suppose first that δ and η are gauge related. Thus, there exists $\theta \in M_0$ such that

$$\eta = e^{\mathrm{ad}_\theta}(\delta) - \frac{e^{\mathrm{ad}_\theta} - 1}{\mathrm{ad}_\theta}(D\theta).$$

As $D\theta = [d, \theta]$,

$$\frac{e^{\mathrm{ad}_\theta} - 1}{\mathrm{ad}_\theta}(D\theta) = \sum_{i\geq 0} \frac{\mathrm{ad}_\theta^i}{(i+1)!}[d, \theta] = -\sum_{i\geq 1} \frac{\mathrm{ad}_\theta^i}{i!} d.$$

Therefore,

$$d+\eta = d + e^{\mathrm{ad}_\theta}(\delta) + \sum_{i\geq 1}\frac{\mathrm{ad}_\theta^i}{i!}(d) = e^{\mathrm{ad}_\theta}(d+\delta).$$

We then use the general formula $e^{\mathrm{ad}_\theta}(d+\delta) = e^\theta(d+\delta)e^{-\theta}$ (see for instance [5, Proposition 4.13]) to conclude that

$$d+\eta = e^\theta(d+\delta)e^{-\theta}, \text{ that is, } (d+\eta)e^\theta = e^\theta(d+\delta),$$

and e^θ is the required isomorphism.

For the other implication simply reverse the above argument. □

Due to this fact we often identify M_0 with

$$\exp(M_0) = \{e^\theta,\ \theta \in M_0\}, \tag{12}$$

and denote $\widetilde{\mathrm{MC}}(M) = \mathrm{MC}(M)/\exp(M_0)$.

Now assume that M is of finite type and choose basis $\{\partial_i\}_{i=1}^s$ and $\{\sigma_\ell\}_{\ell=1}^r$ of M_{-1} and M_{-2} respectively. Write

$$[\partial_i,\partial_j] = \sum_\ell \lambda_{ij}^\ell\,\sigma_\ell,\quad \lambda_{ij}^\ell \in \mathbb{Q}.$$

Then, a straightforward computation shows that, an arbitrary derivation $d+\delta = \sum_i \alpha_i\partial_i$ of degree -1 is a differential if and only if

$$\sum_{i,j}\lambda_{ij}^\ell\,\alpha_i\alpha_j = 0,\quad \ell = 1,\dots,r.$$

In other words, if we denote by $\mathbf{V}_L \subset \mathbb{C}^s$ the affine algebraic variety defined by the polynomials $\sum_{i,j}\lambda_{ij}^\ell\,\alpha_i\alpha_j$, with $\ell = 1,\dots,r$, we conclude that

$$\mathrm{MC}(M) = \{\text{rational points of } \mathbf{V}_L\}. \tag{13}$$

We will now describe the geometrical interpretation of $\widetilde{\mathrm{MC}}(M)$ for certain M as above.

We begin by choosing a connected cgl π. Then, an exact dual procedure to the classical commutative case [15, §3], carried out in the non complete case in [20, Chapter I], let us consider the *bigraded Lie model of* π. In other terms, the minimal Lie model of $(\pi,0)$,

$$\rho\colon (\widehat{\mathbb{L}}(V),d) \xrightarrow{\simeq} (\pi,0),$$

can be chosen so that:

- $V = \oplus_{p,q\geq 0} V_p^q$ is bigraded being the lower grading the usual homological one. This bigradation extends bracket-wise to $\widehat{\mathbb{L}}(V)$.
- $dV^0 = 0$ and $d(V^{n+1}) \subset \widehat{\mathbb{L}}(V^{\leq n})^n$, for $n \geq 0$. In particular d decreases by one the upper degree so that $H(\widehat{\mathbb{L}}(V),d) = \oplus_{p,q\geq 0} H_p^q\big(\widehat{\mathbb{L}}(V,d)\big)$ is also bigraded.

- $\rho\colon \widehat{\mathbb{L}}(V^0) \twoheadrightarrow \pi$ is surjective, $\rho(V^n) = 0$ for $n \geq 1$, $H^0(\rho)\colon H^0(\widehat{\mathbb{L}}(V), d) \xrightarrow{\cong} \pi$ is an isomorphism, and $H^+(\widehat{\mathbb{L}}(V), d) = 0$.

In this bigraded model the elements of $\widehat{\mathbb{L}}(V)^n_p$ are said to have *weight* p-n. Observe that the differential d preserves weight since $dV^n_p \subset \widehat{\mathbb{L}}(V)^{n-1}_{p-1}$. In this setting we consider the sub Lie algebra

$$\mathfrak{Der}\,\widehat{\mathbb{L}}(V) \subset \operatorname{Der}\widehat{\mathbb{L}}(V)$$

of derivations which raise the weight. That is, if $W^m \subset \widehat{\mathbb{L}}(V)$ denotes the subspace of elements of weight m, then $\theta \in \mathfrak{Der}\,\widehat{\mathbb{L}}(V)$ if $\theta(W^m) \subset W^{\geq m+1}$ for all $m \in \mathbb{Z}$. This is a complete dgl for which the following can be proved using a dual argument to the one in [22, Theorem 4.1]:

Theorem 5.2. [9, Corollary 3.6] *The set* $\operatorname{Ho}\mathbf{sset}_\pi$ *of homotopy types of rational CW-complexes whose homotopy Lie algebra is isomorphic to* π *is in bijective correspondence with* $\widetilde{\mathrm{MC}}\big(\mathfrak{Der}\,\widehat{\mathbb{L}}(V), D)\big)$.

In view of this result and the identity (13) we have a bijection

$$\operatorname{Ho}\mathbf{sset}_\pi \cong \mathbf{V}_L/\exp\big(\mathfrak{Der}_0\widehat{\mathbb{L}}(V)\big)$$

exhibiting $\operatorname{Ho}\mathbf{sset}_\pi$ as a moduli space.

Next, we consider H a simply connected graded vector space and denote by $\operatorname{Ho}\mathbf{sset}^1_H$ the class of homotopy types of rational simply connected CW-complexes with reduced homology isomorphic to H. Denote $V = s^{-1}H$ and consider the simply connected dgl $L = (\mathbb{L}(V), 0)$ and the sub dgl $\mathcal{D}\mathrm{er}L$ of $\mathrm{Der}L$ defined by

$$\mathcal{D}\mathrm{er}_k L = \begin{cases} \mathrm{Der}_k L, & \text{if } k > 0,\\ \theta \in \mathrm{Der}_k L, \text{ such that } \theta(V) \subset \widehat{\mathbb{L}}^{\geq 2}(V), & \text{if } k \leq 0. \end{cases}$$

Then we have

Theorem 5.3. [9, Corollary 4.4] *There is an action of* $\operatorname{aut}(V)$ *on* $\widetilde{\mathrm{MC}}(\mathcal{D}\mathrm{er}L)$ *which induces a bijection*

$$\operatorname{Ho}\mathbf{sset}^1_H \cong \widetilde{\mathrm{MC}}(\mathcal{D}\mathrm{er}L)/\operatorname{aut}(V).$$

Sketch of proof. As remarked at the beginning of this section note the MC set of $\mathcal{D}\mathrm{er}L$ consists of decomposable differentials on $\mathbb{L}(V)$. Therefore, the group $\operatorname{aut}(L)$ acts on $\mathrm{MC}(\mathcal{D}\mathrm{er}\,L)$ by

$$\varphi \cdot \delta = \varphi\delta\varphi^{-1}, \quad \varphi \in \operatorname{aut}(L), \quad \delta \in \mathrm{MC}(\mathcal{D}\mathrm{er}\,L). \tag{14}$$

In other words, $\varphi \cdot \delta = \delta'$ if

$$\varphi\colon (\mathbb{L}(V), \delta) \xrightarrow{\cong} (\mathbb{L}(V), \delta')$$

is a dgl isomorphism. Also, it is not hard to see that the map

$$\mathrm{MC}(\mathcal{D}\mathrm{er}\, L) \to \mathrm{Ho}\, \mathbf{sset}^1_H, \qquad \delta \mapsto \langle(\mathbb{L}(V), \delta)\rangle,$$

induces a bijection on the orbit set,

$$\mathrm{MC}(\mathcal{D}\mathrm{er}\, L)/\mathrm{aut}(L) \xrightarrow{\cong} \mathrm{Ho}\, \mathbf{sset}^1_H. \tag{15}$$

On the other hand, see (9), $\exp(\mathcal{D}\mathrm{er}_0 L)$ is the subgroup of $\mathrm{aut}(L)$ consisting of those automorphisms which induce the identity on V and thus, we have the short exact sequence

$$\exp(\mathcal{D}\mathrm{er}_0\, L) \longrightarrow \mathrm{aut}(L) \longrightarrow \mathrm{aut}(V).$$

Observe that, by Proposition 5.1 and with the identification in (12), the action of $\mathrm{aut}(L)$ on $\mathrm{MC}(\mathcal{D}\mathrm{er}L)$ restricts to the gauge action of $\exp(\mathcal{D}\mathrm{er}_0 L)$. Thus V acts on the quotient $\mathrm{MC}(\mathcal{D}\mathrm{er}L)/\exp(\mathcal{D}\mathrm{er}_0 L) = \widetilde{\mathrm{MC}}(\mathcal{D}\mathrm{er}L)$.

Finally, one easily checks that

$$\widetilde{\mathrm{MC}}(\mathcal{D}\mathrm{er}L)/V \cong \mathrm{MC}(\mathcal{D}\mathrm{er}L)/\mathrm{aut}(L) \tag{16}$$

which, by (15), is in bijective correspondence with $\mathrm{Ho}\, \mathbf{sset}^1_H$. □

Observe that, in view of (16) and (13), one exhibits

$$\mathrm{Ho}\, \mathbf{sset}^1_H \cong \mathbf{V}_L/\mathrm{aut}(L)$$

as a moduli space.

To finish we point out that, in the same spirit and with similar arguments, one can describe the set of rational homotopy types sharing the same cohomology algebra: let A be the augmentation ideal of a given simply connected commutative graded algebra of finite type an denote by $\mathrm{Ho}\, \mathbf{sset}^1_A$ the class of homotopy types of rational simply connected CW-complexes with reduced cohomology algebra isomorphic to A.

Let $\mathscr{L}$ denotes the classical Quillen functor from coalgebras to Lie algebras and consider the simply connected dgl $L = \mathscr{L}(A^\sharp)$. Note that, unlike in the previous context, L has a purely quadratic differential, taht is, its differential increases the bracket length exactly by one.

Define $\mathscr{D}\mathrm{er}L$ as the complete sub dgl of $\mathcal{D}\mathrm{er}L$ given by:

$$\mathscr{D}\mathrm{er}_k L = \begin{cases} \mathcal{D}\mathrm{er}_k L, & \text{if } k \geq 0, \\ \{\eta \in \mathrm{Der}_k L, \text{ such that } \eta(V) \subset \widehat{\mathbb{L}}^{\geq 3}(V)\}, & \text{if } k < 0. \end{cases}$$

In other words $\mathscr{D}\mathrm{er}L$ consist of all derivations in positive degrees, all derivations that raise the bracket length in degree zero and those derivations that increase the bracket length by at least 2 in negative degrees. The, $\mathscr{D}\mathrm{er}L$ is a sub cdgl of $\mathrm{Der}L$ for which the following holds:

Theorem 5.4. [10, Theorem 5.3] *There is an action of* $\mathrm{aut}(A)$ *on* $\widetilde{\mathrm{MC}}(\mathscr{D}\mathrm{er}L)$ *which induces a bijection*

$$\widetilde{\mathrm{MC}}(\mathscr{D}\mathrm{er}L)/\mathrm{aut}(A) \cong \mathrm{Ho}\, \mathbf{sset}^1_A.$$

References

1. Baues, H.J., Lemaire, J.M.: Minimal models in homotopy theory. Math. Ann. **225**, 219–242 (1977)
2. Bousfield, A.K., Kan, D.M.: Homotopy Limits, Completions and Localizations. Lecture Notes in Mathematics, vol. 304. Springer, Heidelberg (1972)
3. Buijs, U., Félix, Y., Murillo, A., Tanré, D.: Homotopy theory of complete Lie algebras and Lie models of simplicial sets. J. Topol. **11**, 799–825 (2018)
4. Buijs, U., Félix, Y., Murillo, A., Tanré, D.: Lie models of simplicial sets and representability of the Quillen functor. Isr. J. Math. **238**, 313–358 (2020)
5. Buijs, U., Félix, Y., Murillo, A., Tanré, D.: Lie Models in Topology, Progress in Mathematics, vol. 335. Springer, Cham (2021)
6. Buijs, U., Félix, Y., Murillo, A., Tanré, D.: The infinity Quillen functor, Maurer-Cartan elements and DGL realizations. arXiv:1702.04397
7. Dror, E., Zabrodsky, A.: Unipotency and nilpotency in homotopy equivalences. Topology **18**, 187–197 (1979)
8. Félix, Y., Fuentes, M., Murillo, A.: Lie models of homotopy automorphism monoids and classifying fibrations. Adv. in Math. **402**, 1–64 (2022)
9. Félix, Y., Fuentes, M., Murillo, A.: Realization of Lie algebras of derivations and moduli spaces of some rational homotopy types. arXiv:2206.14124
10. Félix, Y., Fuentes, M., Murillo, A.: There is only one realization of Lie algebras. arXiv:2207.10886
11. Félix, Y., Halperin, S., Thomas, J.-C.: Rational Homotopy Theory. Graduate Texts in Mathematics, vol. 205. Springer, New York (2001)
12. Fresse, B.: Homotopy of operads and Grothendieck-Teichmüller groups. Part 1, The algebraic theory and its topological background. Mathematical Surveys and Monographs, vol. 217. American Mathematical Society, Providence, RI (2017)
13. Fuentes, M.: Classification of fibrations with prescribed holonomy action. Topology Appl. **317**, 1–10 (2022)
14. Getzler, E.: Lie theory for nilpotent L_∞ algebras. Ann. of Math. **170**, 271–301 (2009)
15. Halperin, S., Stasheff, J.: Obstructions to homotopy equivalences. Adv. in Math. **32**, 233–279 (1979)
16. Hinich, V.: Descent of Deligne groupoids. Int. Math. Res. Not. **5**, 223–239 (1997)
17. Majewski, M.: Rational homotopical models and uniqueness. Mem. Amer. Math. Soc. **682** (2000)
18. Mal'tsev, A.I.: On a class of homogeneous spaces. Izv. Akad. Nauk SSSR Ser. Mat. **13**, 9–32 (1949)
19. May, J.P.: Classifying spaces and fibrations. Memoirs Am. Math. Soc. **155** (1975)
20. Oukili, A.: Sur l'homologie d'une algébre différentielle, Ph.D. Thesis, Université de Nice (1978)
21. Quillen, D.: Rational homotopy theory. Ann. of Math. **90**, 205–295 (1969)
22. Schlessinger, M., Stasheff, J.: Deformation theory and rational homotopy type. arXiv:1211.1647 (2012)
23. Stasheff, J.D.: A classification theorem for fiber spaces. Topology **2**, 239–246 (1963)
24. Sullivan, D.: Infinitesimal computations in topology. Publ. IHES **47**, 269–331 (1977)
25. Tanré, D.: Homotopie Rationnelle: Modèles de Chen, Quillen, Sullivan. Lecture Notes in Mathematics, vol. 1025. Springer, Cham (1983)

The Calderón Problem for Local and Nonlocal Schrödinger Equations

María Ángeles García-Ferrero(✉)

Departament de Matemàtiques i Informàtica, Universitat de Barcelona, Gran via de les Corts Catalanes 585, 08007 Barcelona, Spain
garciaferrero@ub.edu

Abstract. These notes are based on the talk with the same title given at the RSME Biennial Conference in 2022, based in turn on joint works with Giovanni Covi, Angkana Rüland and Wiktoria Zatoń [7,9]. In them, we survey some results about uniqueness and stability of inverse problems associated with certain local and nonlocal Schrödinger operators. These results are characterized by the exploitation of Runge approximation results and unique continuation properties for the corresponding equations.

1 Introduction

The study of inverse problems is one of the most active fields of modern applied mathematics. Roughly speaking, an inverse problem seeks, from a set of observations, to determine the causal factors that produced them. Among everyday inverse problems, an illustrative example can be the *echolocation*, used by dolphins or bats to locate and identify objects (e.g. potential preys). These mammals emit sounds which are reflected by the obstacles. From these echoes, the animals are able to obtain information about the distance and size of the objects around and act consequently.

Another naive, as well as entertaining, example is the logic puzzle called *nonogram* (see Fig. 1). In this case we have a black-and-white picture on a grid. The observations that we are provided with are the lengths of the coloured blocks in each row and column and the task is to discover the picture behind them.

Inside the context of partial differential equations, many inverse problems deal with the reconstruction of the (properties of the) interior of a medium in a non-invasive and non-destructive way. Several techniques in medicine, industry, geophysics or environmental science are based on this kind of inverse problems. A prototypical example is the *Calderón problem*, which study the determination of the electric conductivity of a medium from voltage and current measurements on its boundary.

This problem is named after the Argentinian mathematician Alberto Calderón, who set the basis of this question in a short article in 1980 [5]. His interest in this problem rose in the 1940's, when he was working as an engineer for an oil company. The Calderón problem, also known as Electrical Impedance Tomography, is not only the basis of geophysical prospecting, but also of medical imaging for the diagnosis of tumours or for monitoring pulmonary functions.

H. Herrero et al. (Eds.): RSME 2022, RSME 13, pp. 81–99, 2024.
https://doi.org/10.1007/978-3-031-62025-6_5

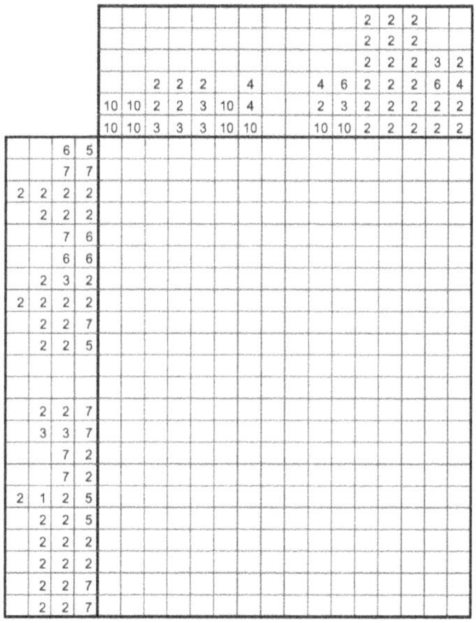

Fig. 1. A nonogram or Japanese puzzle, example of an inverse problem. The grid hides a picture made of black and white squares. The data at the top and in the left are the numbers of consecutive black squares in the corresponding column or row. Each block must be separated by at least a white square.

Since the 1980s, its study has led many developments in the area of inverse problems.

In mathematical terms, the Calderón problem can be formulated as follows: The medium will be a bounded, open set of $\mathbb{R}^n$ denoted by Ω. Unless otherwise stated, we will assume that its boundary $\partial\Omega$ is smooth, although many results require much less regularity (C^2 or even Lipschitz continuous). Assuming that the medium is isotropic, its conductivity will be a bounded, positive function $\gamma : \Omega \to \mathbb{R}$ satisfying $0 < c \leq \gamma(x) \leq C < \infty$ for any $x \in \Omega$. Lastly, $v : \bar{\Omega} \to \mathbb{R}$ will denote the voltage on the medium. Under the absence of sinks or sources of electrical charge, v will obey the following Dirichlet problem:

$$\begin{aligned} \nabla \cdot \big(\gamma(x)\nabla v\big) &= 0 \text{ in } \Omega, \\ v &= g \text{ on } \partial\Omega. \end{aligned} \tag{1}$$

The bulk equation corresponds to the generalization of the Ohm's law and g represents the voltage applied at the boundary.

As previously said, in the Calderón problem we start with the knowledge of current and voltage measurements on the boundary of Ω. This information is encoded in the so-called Dirichlet-to-Neumann map, which maps the voltage imposed on the boundary to the current flux through the boudary:

$$\begin{aligned} \Lambda_\gamma : H^{\frac{1}{2}}(\partial\Omega) &\longrightarrow \big(H^{\frac{1}{2}}(\partial\Omega)\big)^*, \\ g &\longmapsto \gamma\partial_\nu v\big|_{\partial\Omega}. \end{aligned} \tag{2}$$

where ∂_ν is the normal derivative and v is the (unique) solution to (1). We notice on one hand that this operator is well-defined as a map between the given function spaces[1], due to the well-posedness of the direct problem (1) for bounded, strictly positive functions γ. On the other hand, the Neumann data $\gamma\partial_\nu v\big|_{\partial\Omega}$ must be understood in general in a weak sense[2]. Lastly, we remark that in practice, it is the current flux through the boundary what is prescribed and the generated voltage on the boundary what is measured.

Using the previous notation, we can now refer to the Calderón problem in very simple terms as the study of the map

$$\Lambda_\gamma \longrightarrow \gamma. \tag{3}$$

The main questions concerning the Calderón problem, and inverse problems in general, are those of uniqueness, stability and reconstruction. In our case and roughly speaking:

- *Uniqueness:* Does $\Lambda_{\gamma_1} = \Lambda_{\gamma_2}$ imply $\gamma_1 = \gamma_2$?
- *Stability:* If Λ_{γ_1} and Λ_{γ_2} are close, are then γ_1 and γ_2 also close?
- *Reconstruction:* Is there a constructive algorithm to obtain γ from Λ_γ?

Some other crucial questions concern more realistic situations. For instance, what happens in the case that we have access only to a subset $\Gamma \subset \partial\Omega$ of the boundary? In this situation, we will talk about the *partial-data* Calderón problem.

During the 1980 s, fundamental contributions by Calderón, Sylvester, Uhlmann, Nachman and Alessandrini [1,5,23,27] answered the main questions for sufficiently regular conductivities and dimension $n \geq 3$. In this case, the conductivity equation can be rewritten as a Schrödinger equation. This is the setting we are interested in and a little bit more will be said on this matter in Sect.1.1. Further main contributions could be mentioned, including those in the 2000 s on the two-dimensional problem [2,3] or more recent results considering the problem in three or higher dimensions but with less regular conductivities [6,12]. In addition, the anisotropic Calderón problem would deserve to be introduced. Nevertheless, all these things lie outside the scope of this exposition and great surveys or introductory notes on the topics can be found by the interested reader (see e.g. [8,29,30]).

1.1 The Calderón Problem for the Schrödinger Equation

In order to study the Calderón problem for the conductivity equation (3) for sufficiently regular conductivities (namely $\gamma \in C^2(\bar{\Omega})$), this is reduced to an

1 $H^{\frac{1}{2}}(\partial\Omega) = \{\phi|_{\partial\Omega} : \phi \in H^1(\Omega)\}$.

2 For any $h \in H^{\frac{1}{2}}(\partial\Omega)$, $\langle \Lambda_\gamma f, h\rangle = \int_\Omega \gamma\nabla v \cdot \nabla\tilde{h}\, dx$, where $\tilde{h} \in H^1(\Omega) : \tilde{h}|_{\partial\Omega} = h$..

analog inverse problem for a Schrödinger equation. Under the identification

$$u = \gamma^{\frac{1}{2}} v, \quad q = \frac{\Delta \gamma^{\frac{1}{2}}}{\gamma^{\frac{1}{2}}}, \quad f = (\gamma^{\frac{1}{2}} g)\big|_{\partial\Omega},$$

the conductivity problem (1) can be rewritten as

$$\begin{aligned} (-\Delta + q)u &= 0 \text{ in } \Omega, \\ u &= f \text{ on } \partial\Omega. \end{aligned} \tag{4}$$

The information about boundary measurements in this problem is again encoded in an analog Dirichlet-to-Neumann map:

$$\begin{aligned} \Lambda_q : H^{\frac{1}{2}}(\partial\Omega) &\longrightarrow \left(H^{\frac{1}{2}}(\partial\Omega)\right)^* \\ f &\longmapsto \partial_\nu u\big|_{\partial\Omega} \end{aligned} \tag{5}$$

where u is given by the solution of (4). As we remarked above for (2), the Dirichlet-to-Neumann map Λ_q is well-defined provided the problem (4) is well-posed. In the case that $q \in L^\infty(\Omega)$ is related with γ as above, this holds. In general, we consider $q \in L^\infty(\Omega)$ such that 0 is not an eigenvalue of $-\Delta + q$ in Ω. Under this assumption, we will be concerned with the map

$$\Lambda_q \longrightarrow q, \tag{6}$$

which corresponds to the Calderón problem for the Schrödinger equation. The uniqueness, (in-)stability and reconstruction of this problem (in $n \geq 3$) were tackled in [1,23,27], respectively. Before sketching some ideas about the first two points, we remark that in order to apply the results for (6) to the original Calderón problem (3), boundary determination is required [18,28] so, in particular, Λ_q is known from Λ_γ.

The proof of uniqueness is based on the so-called Alessandrini's identity:

$$\int_\Omega (q_1 - q_2) u_1 u_2 dx = \int_{\partial\Omega} (\Lambda_{q_1} - \Lambda_{q_2}) f_1 f_2 d\sigma, \tag{7}$$

where u_i, $i \in \{1, 2\}$, are the (unique) solutions to

$$\begin{aligned} (-\Delta + q_i)u_i &= 0 \text{ in } \Omega, \\ u_i &= f_i \text{ on } \partial\Omega. \end{aligned}$$

$\Lambda_{q_1} = \Lambda_{q_2}$ implies that the right-hand side of (7) vanishes. Therefore, in order to conclude that $q_1 = q_2$ we need to be aware of the existence of solutions u_i such that the set of products $\{u_1 u_2\}$ is dense.

These special solutions were constructed by Sylvester and Uhlmann in [27] and are known as complex geometric optics (CGO) solutions. Broadly speaking, they are perturbations of harmonic functions of the form $e^{\zeta \cdot x}$ with $\zeta \in \mathbb{C}^n$ and

$\zeta \cdot \zeta = 0$. That is, given $q \in L^\infty(\Omega)$, there are solutions to $(-\Delta + q)u = 0$ (in $\mathbb{R}^n$, where q must be understood as its extension by 0 outside Ω) of the form

$$u(x) = e^{\zeta \cdot x}\big(1 + r_{\zeta,q}(x)\big)$$

with $r_{\zeta,q} \to 0$ (in certain sense) as $|\zeta| \to \infty$. Taking ζ_1, ζ_2 suitably so $\zeta_1 \cdot \zeta_2 = i\xi$ for any $\xi \in \mathbb{R}^n$ and $|\zeta_i| \to \infty$, we end up with $\int_{\mathbb{R}^n}(q_1 - q_2)e^{i\xi\cdot x}dx = 0$ for all $\xi \in \mathbb{R}^n$.

The stability question for (6) is equivalent to the continuity of the inverse of the (continuous, injective) operator $q \in L^\infty(\Omega) \mapsto \Lambda_q \in B\big(H^{\frac{1}{2}}(\partial\Omega), (H^{\frac{1}{2}}(\partial\Omega))^*\big)$. If we consider the restriction of this operator to a compact function set of potentials, a continuous estimate for its inverse holds [17, Lemma 2.1]. More precisely, assuming $\|q\|_{L^\infty(\Omega)} \leq M$, there exists a modulus of continuity η[3] such that

$$\|q_1 - q_2\|_{H^{-1}(\Omega)} \leq \eta(\|\Lambda_{q_1} - \Lambda_{q_2}\|_*, \tag{8}$$

where the notation $\|\cdot\|_*$ refers to the norm as a map between the specified function spaces. Using again CGO solutions, Alessandrini proved in [1] that $\eta(t) \leq C(-\log t)^{-\nu}$ for t sufficiently small. This logarithmic dependence, which is shown to be optimal [22], implies that the size of the error in the determination of q decreases much slower than the error in measuring Λ_q. This is troublesome for applications and therefore stability improvements under extra hypothesis are necessary.

In the remaining of this article we will deal with the partial-data Calderón problem (6) in the following two situations:

- For the Helmholtz acoustic equation, where q is replaced by $-qk^2 + V$ with q positive and k playing the role of the frequency. Obviously, the previous results apply in this case. Nevertheless, combining the CGO solutions with suitable quantitative Runge-type approximation results, we are able to obtain improved stability results when the parameter k increases.
- The Laplacian is replaced by a nonlocal operator. First of all, these cases require to define the analogs to (4) and (5) properly. Secondly, we will see that (directionally) antilocal conditions, kind of global versions of weak unique continuation, and the Runge-type results we can infer from them, allow to prove uniqueness for the associated partial-data inverse problem. We will see that even one single measurement may be enough to determine the potential q uniquely.

1.2 Outline of the Article and Disclaimer

The rest of the article is structured as follows: In Sect. 2 we introduce the so-called Runge approximation results for elliptic local equations and we see a quantitative version for the acoustic Helmholtz equation. In addition, we illustrate

[3] $\eta : [0, \infty) \to [0, \infty)$ is continuous, non-decreasing and $\eta(t) \to 0$ as $t \to 0$.

the significance of these results with the improvements in the stability of the partial-data Calderón problem for the acoustic Helmholtz equation [9]. Section 3 begins with an introduction to the fractional counterpart of the Calderón problem (6), which appeared firstly in [11], and the key property behind uniqueness results for it: the antilocality condition of the fractional Laplacian. We then continue considering nonlocal operators which take into account how the function behaves in conical domains and for which nothing more than antilocality in cones can be expected. We see the implications of this directionally antilocality in the approximation theorems and in the associated Calderón problem [7].

These notes are thought, as the talk was, for a general audience. Therefore, we will neither be exhaustive nor too technical, but we will focus on sharing the flavour of some results. Proofs will not be provided and the interested reader is kindly referred to the original sources in order to find complete details. In addition, we will make use of simple pictures to show the settings of the results here presented.

2 Runge Approximation and the Calderón Problem for the Acoustic Helmholtz Equation

In this section we are interested in the acoustic Helmholtz operator

$$P_k = -\Delta - k^2 q + V, \tag{9}$$

where q is a strictly positive function on Ω and k plays the role of the frequency, and in the associated partial-data Calderón problem

$$\Lambda_{q,V}(k) = \Lambda_{-k^2 q+V} \longrightarrow \{q, V\}. \tag{10}$$

It is clear that the results mentioned in the introduction about uniqueness, (conditional, logarithmic) stability and reconstruction for (6) apply in this case, just replacing q by $-k^2 q + V$. Nevertheless, our interest lies in the stabilizing effect of high frequencies.

Stability improvements with increasing frequencies have been already observed, for instance for the acoustic or the Helmholtz equation and full data [13,14,24] and for the Helmholtz equation and partial data with impedance boundary conditions [19]. In [9] we deal with the Calderón problem (10) with partial data, i.e. if we know $\Lambda_{q,V}(k)$ only on a subset $\Gamma \subset \partial\Omega$. Our result exploits quantitative Runge-type approximation theorems that we obtain with explicit dependence on the frequency parameter k.

2.1 Runge Approximation Results

Let $\Omega, \Omega' \subset \mathbb{R}^n$ be bounded, open sets with $\Omega' \Subset \Omega$ and let P a differential operator. We will say that P satisfies a *Runge-type approximation result* if the following holds:

For any solution v of $Pv = 0$ in Ω', there exists a solution u of $Pu = 0$ in Ω such that the difference $v - u|_{\Omega'}$ can be arbitrarily small (in a suitable norm). (11)

These results are named after Runge theorem of 1884 [26] in complex analysis, which states that any holomorphic function in $\Omega' \subset \mathbb{C}$ can be approximated by holomorphic functions in $\Omega \supset \Omega'$ provided Ω' does not have holes in Ω, i.e. $\Omega \backslash \Omega'$ is connected.

The idea of Runge theorem was resumed in the 1950 s by Lax and Malgrange [20,21] and later generalized by Browder [4]). They proved that some classes of elliptic operators satisfy a Runge-type approximation result (11) provided that $\Omega \backslash \Omega'$ is connected (see Fig. 2).

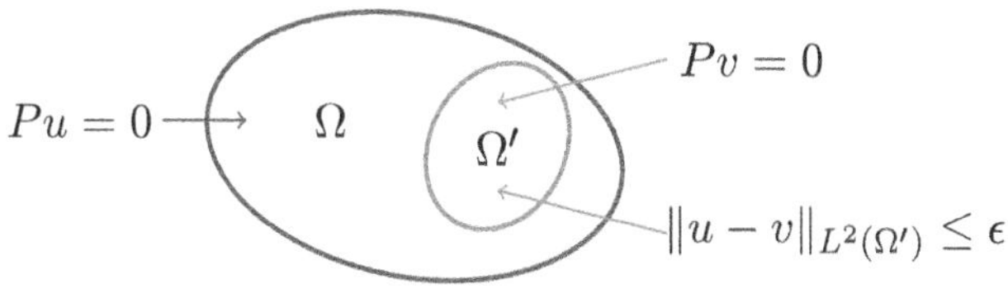

Fig. 2. Illustrative setting for the Runge approximation result for an elliptic operator P, such as the Laplacian.

Runge-type approximation results can be inferred from *(weak) unique continuation properties* for (the adjoint of) the corresponding operator. These properties are a generalization of analytic continuation and read as follows (for more information, see e.g. [31]):

$$\textit{If } Pu = 0 \textit{ in } \Omega \textit{ and } u = 0 \textit{ in } \Omega', \textit{ then } u = 0 \textit{ in } \Omega. \tag{12}$$

We could instead consider the following version of unique continuation from local Cauchy data:

$$\textit{If } Pu = 0 \textit{ in } \Omega \textit{ and } u|_\Gamma = 0 = \partial_\nu u|_\Gamma;^4 \textit{ then } u = 0 \textit{ in } \Omega. \tag{13}$$

[4] Here and from now on, $\Gamma \subset \partial\Omega$ is assumed to be a non-empty, relatively open subset. Using similar duality arguments, this unique continuation leads to the following version of Runge-type approximation, in which further conditions on the approximating solution are considered:

For any solution v of $Pv = 0$ in Ω', there exists a solution u of $Pu = 0$ in Ω with $\text{supp}(u|_{\partial\Omega}) \subseteq \Gamma$ *and such that the difference $v - u|_{\Omega'}$ can be arbitrarily small (in a suitable norm).* (14)

Both (qualitative) statements (13) and (14) can be made *quantitative.* In the former, this means that we quantify how close two solutions are in terms of the closeness of its boundary values on Γ. In the latter, we quantify how we can improve the approximation at the expense of admitting that u may behave more

[4] Here we are assuming P is a second-order elliptic operator. In the case of higher-order operators, we would have to consider higher-order boundary conditions.

poorly on Γ. This kind of result is well-illustrated in the following theorem (see also Fig. 3) for the operator P_k in (9). Our main contribution in [9] is the explicit dependence on the parameter k:

Theorem 2.1 ([9], Theorem 1.1). *Let $n \geq 2$ and let $\Omega, \Omega' \subset \mathbb{R}^n$ be bounded, connected, Lipschitz open sets such that $\Omega' \Subset \Omega$ and $\Omega \backslash \Omega'$ is connected. Let $\Gamma \subset \partial\Omega$ be a non-empty, relatively open subset. Let $q \in C^{0,1}(\Omega)$ be strictly positive and let $V \in L^\infty(\Omega)$ satisfy some mild conditions.*[5] *Then for any solution $v \in H^1(\Omega')$ of $P_k v = 0$ in Ω', with $k \geq 1$ verifying additional technical conditions(See footnote 5) and for any $\epsilon > 0$, there exists a solution $u \in H^1(\Omega)$ of $P_k u = 0$ in Ω with $\operatorname{supp}(u|_{\partial\Omega}) \subseteq \Gamma$ and such that*

$$\|u - v\|_{L^2(\Omega')} \leq \epsilon \|v\|_{H^1(\Omega')}, \qquad \|u\|_{H^{1/2}(\partial\Omega)} \leq Ce^{Ck^s \epsilon^{-\mu}} \|v\|_{L^2(\Omega')},$$

where C, s and μ are constants independent of k, v and ϵ.

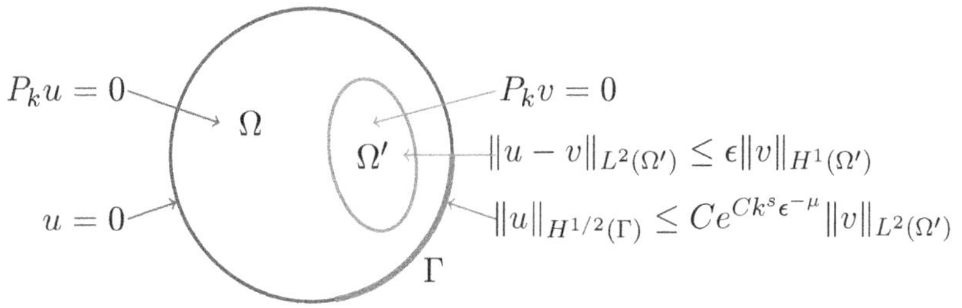

Fig. 3. Graphical statement of Theorem 2.1.

The proof of Theorem 2.1 is based on duality arguments and unique continuation estimates in which we track the dependence on k:

Proposition 2.2 ([9], Proposition 2). *Under the same hypothesis of Theorem 2.1, let $u \in H^1(\Omega)$ such that $P_k u = 0$ in Ω and*

$$\|u\|_{H^1(\Omega)} \leq M, \quad \|u\|_{H^{\frac{1}{2}}(\Gamma)} + \|\partial_\nu u\|_{H^{-\frac{1}{2}}(\Gamma)} \leq \epsilon.$$

Then

$$\|u\|_{L^2(\Omega)} \leq Ck \left| \log\left(\frac{\epsilon}{M + \epsilon}\right) \right|^{-\mu} (M + \epsilon),$$

where C and μ are independent of M, ϵ and k.

[5] We assume 0 is not a Dirichlet eigenvalue of $-\Delta + V$ in Ω. Then we can infer that the operator $(-\Delta + V)^{-1} M_q$, where M_q denotes the multiplication operator with q, has a discrete set of eigenvalues $\Sigma_{q,V}$. We further assume that dist $(k^2, \Sigma_{q,V}) > ck^{2-n}$ with c small. This condition ensures the well-posedness of the direct problem and offers some a priori estimates which are independent of k.

Notice that, similar to what we have seen about the stability of the Calderón problem, estimating the size of u in Ω in terms of the smallness of u on Γ requires some a priori information, namely that its H^1 norm in Ω is bounded by a constant M.

Remark 2.3. *If in Theorem 2.1 we consider that v solves the equation $P_k v = 0$ in a subset $\tilde{\Omega}' \subset \Omega$ which is slightly bigger than Ω' (where the approximation is required), the dependence on ϵ of the estimate for the boundary norm of u improves (from exponential to polynomial, being this optimal). For the complete statement, the reader is referred to [9, Theorem 1.2]. Here, we only show it graphically in Figure 4. We finish noticing that this change in the ϵ-dependence is inherited by improved unique continuation estimates in the interior of Ω. Indeed, under the hypothesis of Proposition 2.2, we infer*

$$\|u\|_{L^2(\Omega')} \leq Ce^{Ck}\epsilon^{\nu}(M+\epsilon)^{1-\nu}.$$

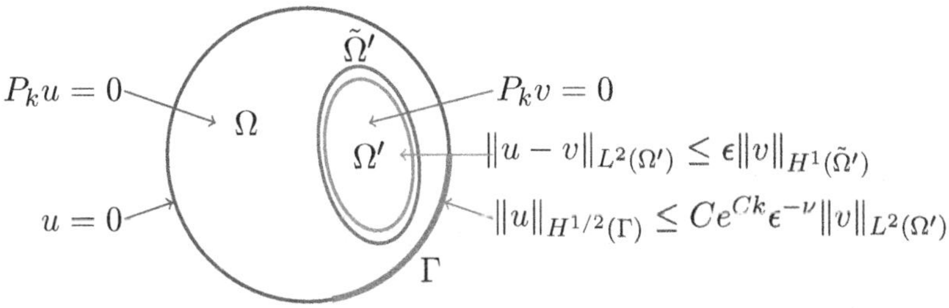

Fig. 4. Graphical statement of the quantitative Runge approximation result with improved ϵ-dependence of Remark 2.3.

2.2 Stability Improvement for the Partial-Data Calderón Problem

In this section, we introduce improved stability estimates for the partial-data Calderón problem for the operator P_k in (9). More precisely, we are interested in the inverse problem

$$\Lambda^{\Gamma}_{q,V}(k) \longrightarrow \{q, V\}. \tag{15}$$

The Dirichlet-to-Neumann map on $\Gamma \subset \partial\Omega$ is given by

$$\Lambda^{\Gamma}_{q,V}(k) : \tilde{H}^{\frac{1}{2}}(\Gamma) \longrightarrow \left(\tilde{H}^{\frac{1}{2}}(\Gamma)\right)^*, \quad f \longmapsto \partial_\nu u\big|_{\partial\Omega},$$

[6] where u is the solution to

$$P_k u = 0 \text{ in } \Omega, \qquad u = f \text{ on } \partial\Omega.$$

[6] $\tilde{H}^{\frac{1}{2}}(\Gamma) = \{f \in H^{\frac{1}{2}}(\partial\Omega) : \text{supp}(f) \subseteq \Gamma\}$.

Given f, the solution u is unique provided we assume some solvability conditions on q, V and k(See footnote 5) Then, $\Lambda^{\Gamma}_{q,V}(k)$ is well-defined (and bounded).

Exploiting the quantitative Runge approximation result from Remark 2.3, we obtain the following stability result for (15):

Theorem 2.4. ([9], Proposition 1). *Let $n \geq 3$, let $\Omega \subset \mathbb{R}^n$ be a bounded, connected, smooth open set and let $\Gamma \subset \partial\Omega$ be a non-empty, relatively open subset. Let $q_1, q_2 \in C^{0,1}(\Omega)$ be strictly positive and $V_1, V_2 \in L^\infty(\Omega)$ satisfy some mild conditions(See footnote 5) and be such that*

$$\|q_j\|_{L^\infty(\Omega)} + \|V_j\|_{L^\infty(\Omega)} \leq B,$$
$$q_1 = q_2,\ V_1 = V_2 \text{ in a neighborhood of } \partial\Omega.$$

Then, for all $k \geq 1$ satisfying additional technical conditions but arbitrarily large(See footnote 5) and for $\delta = \|\Lambda^{\Gamma}_{V_1,q_1}(k) - \Lambda^{\Gamma}_{V_2,q_2}(k)\|_* < 1$, we have

$$\|k^2(q_2 - q_1) - (V_2 - V_1)\|_{H^{-1}(\Omega)} \leq C \left(e^{Ck^{n+3}}\delta + \frac{1}{\left(k + |\log\delta|^{\frac{1}{n+3}}\right)^{\frac{2}{n}}} \right),$$

where C is independent of k and δ.

We firstly notice that the estimate holds under a priori knowledge about the potential in a neighbourhood of boundary $\partial\Omega$. We also emphasize that the stability estimate has a Lipschitz and a logarithmic dependence on the measuring error δ, but the former dominates to the latter when k increases.

The proof of this theorem uses the Alessandrini's identity (7) and suitably chosen CGO solutions. As in the case of uniqueness, this imposes the restriction that $n \geq 3$. Finally, we apply the Runge approximation result of Remark 2.3 (see also Figure 4) to the CGO solutions. After suitably optimization of parameters, the result holds.

3 The Calderón Problem for Nonlocal Operators

In this section we are going to consider the Calderón problem for nonlocal equations. We will start introducing the fractional counterpart of (6), in which the Schrödinger equation is replaced by the fractional version $((-\Delta)^s + q)u = 0$, with $s \in (0, 1)$ [11]. Subsequently, we will consider generalizations in which instead of the fractional Laplacian (which, roughly speaking, "sees" the function u everywhere) we have operators which only "see" (unbounded) cones [7]. In both cases, we will obtain uniqueness results for the corresponding (partial-data) inverse problems relying on ("nonlocal versions" of) unique continuation properties.

3.1 The Fractional Calderón Problem

The fractional counterpart of the Calderón problem (6) was introduced by Ghosh, Salo and Uhlmann in [11]. First of all, the Dirichlet problem turns into

$$\begin{aligned} ((-\Delta)^s + q)u &= 0 \text{ in } \Omega, \\ u &= f \text{ in } \Omega_e. \end{aligned} \tag{16}$$

where $s \in (0,1)$ and $\Omega_e = \mathbb{R}^n \backslash \bar{\Omega}$. We recall that the fractional Laplacian $(-\Delta)^s$ is defined by

$$(-\Delta)^s u = \mathcal{F}^{-1}(|\cdot|^{2s}\mathcal{F}u), \quad u \in H^s(\mathbb{R}^n)$$

[7] where $\mathcal{F}$ denotes the Fourier transform. We can say that fractional Laplacian "sees" the whole space, meaning that $(-\Delta)^s u(x)$ depends on u at any point in $\mathbb{R}^n$. Therefore, in order to define the Dirichlet problem, Cauchy data is required in the exterior domain Ω_e instead of on the boundary.

The analog to the Dirichlet-to-Neumann in (5) is in turn given by

$$\begin{aligned} \Lambda_q^s : H^s(\Omega_e) &\longrightarrow (H^s(\Omega_e))^*, \\ f &\longmapsto (-\Delta)^s u|_{\Omega_e}, \end{aligned} \tag{17}$$

[8] where u solves (16). Once more, this operator is well-defined provided the problem (16) is well-posed, what happens if 0 does not belong to the countable set of eigenvalues of $(-\Delta)^s + q$ in Ω (see [11, Lemma 2.3]). In addition, $\Lambda_q^s f$ must be understood in general in the weak sense, namely, for any $h \in H^s(\Omega_e)$

$$\langle \Lambda_q^s f, h\rangle = \int_{\mathbb{R}^n} ((-\Delta)^{\frac{s}{2}} u)((-\Delta)^{\frac{s}{2}} \tilde{h}) dx \ \text{ where } \tilde{h} \in H^s(\mathbb{R}^n) : \tilde{h}|_{\Omega_e} = h.$$

This is consistent with the bilinear form associated with the notion of (weak) solution of (16).

The main questions about the Calderón problem

$$\Lambda_q^s \longmapsto q \tag{18}$$

have been answered in [10,11,25], where important differences with respect to the local case (6) are observed. Firstly, uniqueness holds from partial data [11]: Let V and W be open subsets of Ω_e (see Figure 5), then

$$\Lambda_{q_1}^s f|_V = \Lambda_{q_2}^s f|_V \text{ for any } f \in C_c^\infty(W) \text{ implies that } q_1 = q_2. \tag{19}$$

Moreover, for regular enough potentials ($C^0(\Omega)$), q can be (constructively) determined by a single measurement [10]: Taking some non-trivial $f \in C_c^\infty(W)$ and knowing $\Lambda_q^s f|_V$, the potential q can be reconstructed.

---

[7] $H^r(\mathbb{R}^n) := \{v \in \mathcal{S}'(\mathbb{R}^n) : \|(1+|\cdot|^2)^{\frac{r}{2}}\mathcal{F}v\|_{L^2(\mathbb{R}^n)} < \infty\}$ for any $r \in \mathbb{R}$.

[8] $H^r(V) := \{v|_V : v \in H^r(\mathbb{R}^n)\}$., equipped with the quotient norm.

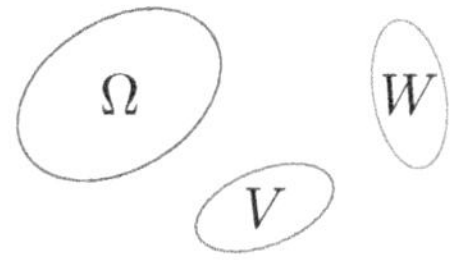

Fig. 5. Setting for the partial-data Calderón problem (18): the Dirichlet-to-Neumann map $\Lambda_q^s f$ is known on the subset V for exterior data f supported in W. (V and W could intersect and even be equal.)

The idea of the proof of (19) is again based in the Alessandrini's formula, which in this setting reads as follows:

$$\int_{\Omega}(q_1 - q_2)u_1 u_2 dx = \int_{\Omega_e}(\Lambda_{q_1}^s - \Lambda_{q_2}^s)f_1 f_2 dx,$$

where u_i solve (16) with potential q_i and exterior data f_i. If we consider that f_1 is supported in W and f_2, in V, then the hypothesis in (19) implies $\int_{\Omega}(q_1 - q_2)u_1 u_2 dx = 0$. Then, the conclusion $q_1 = q_2$ will follow if we have solutions u_i as above such that $\{u_1 u_2\}$ is dense. In this case, this can be directly infer from the following Runge-type approximation result (see [11] and references there) for any $U \subset \Omega_e$:

For any function $v \in \tilde{H}^s(\Omega)$ there exists a solution of (16) with $\mathrm{supp}(f) \subseteq U$ such that $v - u|_{\Omega}$ can be arbitrarily small. (20)

[9] We notice that, contrary to the density result (11) for the Laplacian or similar elliptic local operators, we can approximate any function, and not only solutions, by solutions of the fractional Schrödinger equation (16).

This density result can be seen as a dual property of the *antilocality* of the fractional Laplacian:

If $u = 0 = (-\Delta)^s u$ in U, then $u = 0$ in $\mathbb{R}^n$.

Antilocality is a global unique continuation property, since the conclusion $u = 0$ holds beyond the domain where the equation is satisfies (indeed in the whole $\mathbb{R}^n$). Notice that the (weak) unique continuation property (12) for local operators requires the validity of the equation in the region where we can conclude that the function vanishes. Nevertheless, nonlocal operators which do not "see" the whole space cannot be antilocal and only *directional antilocality* can be expected.

Let us think for instance in the operator

$$Au(x) = \int_0^{\infty} \big(u(x) - u(x+y)\big)\frac{1}{y^{1+2s}}dy.$$

It is clear that the value of $Au(x)$ is independent of how u looks like at the left of x. So any non-trivial function supported at the left of some $x_0 \in \mathbb{R}$ verifies

[9] $\tilde{H}^r(\Omega)$ is the closure of $C_c^\infty(\Omega)$ in $H^r(\mathbb{R}^n)$.

that $Au = 0 = u$ in $(x_0, +\infty)$. Therefore, the operator A is not antilocal, but it is *antilocal to the right* [15]:

If $u = 0 = Au$ in (a, b), then $u = 0$ in $(a, +\infty)$.

In the next section we will consider operators which "see" cones, meaning that the value of the operator applied to some function and evaluated at some point depends on the function in a conical domain with origin at the given point. We will obtain uniqueness results for the associated (partial-data) Calderón problems under the assumption that they satisfy the condition of *antilocality in cones.*

3.2 The Calderón Problem for Antilocal-In-Cones Operators

During this section, $\mathcal{C} \subset \mathbb{R}^n$ will denote a convex, non-empty, open cone. We can keep in mind the example in $\mathbb{R}^2$ of Fig. 6 for simplicity. Given a (bounded, open) set Ω in $\mathbb{R}^n$, we will denote by $\Omega + \mathcal{C}$ the points of the form $x + y$ with $x \in \Omega$ and $y \in \mathcal{C}$ (see also Fig. 6).

In [16], Ishikawa introduced the condition of *antilocality in cones*, which generalizes the above idea of antilocality to the right. We will say that an operator P is antilocal in $\mathcal{C}$ or $\mathcal{C}$-*antilocal* if the following holds:

If $u = 0 = Pu$ in Ω, then $u = 0$ in $\Omega + \mathcal{C}$.

Using the notation $\mathcal{C}^2 = (-\mathcal{C}) \cup (\mathcal{C})$, we may also talk about $\mathcal{C}^2$-antilocal operators. In this case, the fact that $u = 0 = Pu$ in Ω would imply that $u = 0$ in $\Omega + \mathcal{C}^2$ (see Figure 6). In the remaining of this exposition, we are going to see the implications of these directionally-antilocal conditions in density results and in the study of associated Calderón problems.

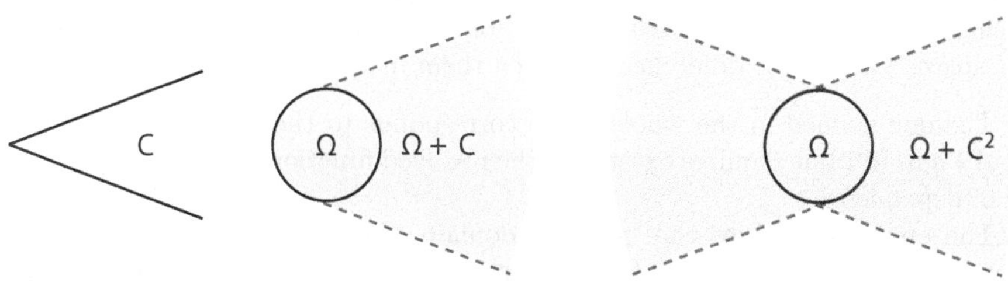

Fig. 6. Illustration of a cone $\mathcal{C}$ and of the subsets $\Omega + \mathcal{C}$ and $\Omega + \mathcal{C}^2$.

A $\mathcal{C}$-antilocal operator must capture how the function is at least in the cone $\mathcal{C}$ with origin at the point of evaluation. Otherwise, we can construct simple counterexamples as in Section 3.1 for antilocality to the right. Here, we will consider operators which "see" one single cone $\mathcal{C}$ or a double cone $-\mathcal{C}$ and $\mathcal{C}$.

Simple examples to think about are the following:

$$\begin{aligned} Au(x) &= \int_C \big(u(x) - u(x+y)\big)\frac{1}{|y|^{n+2s}}dy, \qquad s \in (0, \frac{1}{2}); \\ Su(x) &= \int_C \big(2u(x) - u(x+y) - u(x-y)\big)\frac{1}{|y|^{n+2s}}dy. \end{aligned} \tag{21}$$

Notice that indeed $Au(x)$ and $Su(x)$ depend on u in $\{x\} + \mathcal{C}$ and $\{x\} + \mathcal{C}^2$, respectively. These operators are instances of elliptic nonlocal operators which are generators of $2s$-stable Lévy processes. In the article [7] more general operators of this type are considered and the well-posedness of the associated Dirichlet problems is studied. Furthermore, weaker antilocality-in-cones conditions are obtained for particular cases. In these notes, we will skip these parts and we will only keep in mind the domain dependence of the operators. Therefore, and for the sake of simplicity, we will use the notation A and S to refer to operators "seeing" $\mathcal{C}$ and $\mathcal{C}^2$, respectively.

We shall start by considering the analog to (16) for an operator seeing $\mathcal{C}$. Taking into account the domain of dependence, it is natural to consider the Dirichlet problem for an operator A on the set $\Omega + \mathcal{C}$ and not on the whole space $\mathbb{R}^n$. Indeed, whatever we may prescribe outside the set $\bar{\Omega} + \mathcal{C}$ will not have any effect in the solution of the problem in Ω. Then, the Dirichlet problem we are interested in is

$$\begin{aligned} (A+q)u &= 0 \text{ in } \Omega, \\ u &= f \text{ in } \Omega_{e,\mathcal{C}}, \end{aligned} \tag{22}$$

where $\Omega_{e,\mathcal{C}} = (\Omega + \mathcal{C})\backslash\bar{\Omega}$. In order to define a weak solution of this problem, and subsequently the associated Dirichlet-to-Neumann map, it is necessary to consider an appropriate bilinear form. In this case, we can define the bilinear form in $\mathbb{R}^n \times \mathbb{R}^n$ or restricted to the "visible" domain $(\Omega + \mathcal{C}) \times (\Omega + \mathcal{C})$. Both of them give rise to the same weak solution for a given exterior data in the suitable H^s space[10]. However, differences between them include:

- The one defined in the whole space corresponds to the weak formulation of $A + q$ in $\mathbb{R}^n$, but requires extending the involved functions beyond the domain of dependence.
- The one restricted to the "visible" domain of the operator is well defined in greater, less symmetric function spaces. However, outside Ω, it does not always agrees with the operator A (in the weak sense).

Due to the previous remarks, here we will consider the Dirichlet-to-Neumann map $\Lambda^s_{\mathcal{C},q}$ given by the first bilinear form. In this case, like in (5) and (17), it will correspond to Au (in the weak sense) outside Ω, where u is the solution to

[10] In the case of S, this space is nothing else than $H^s(\Omega_{e,\mathcal{C}^2})$. However, for A, we observe that the suitable space corresponds to the quotient space $H^s(\mathbb{R}^n)/\big(\tilde{H}^s(\Omega) \oplus \tilde{H}^s(\mathbb{R}^n\backslash(\bar{\Omega} + \Gamma))\big)$, which agrees with $H^s(\Omega_{e,\mathcal{C}})$ if $s < \frac{1}{2}$ but which also contains "trace information" on $\partial\Omega\backslash(\Omega + \mathcal{C})$ if $s > \frac{1}{2}$. See more in [7, Lemma 5.13].

(22). Once more, we will work under the assumption that (22) is well-posed[11] so $\Lambda^s_{\mathcal{C},q}$ is well-defined. In order to capture information about the unknown q, we need to measure Au in a region where Ω is seen, i.e. in $\Omega - \mathcal{C}$. Indeed, if we consider $Au|_{\Omega_{e,\mathcal{C}}}$, all the information about u which will be "catched" is about the exterior data f. Therefore, we will consider $Au|_{\Omega_{e,-\mathcal{C}}}$. However, this quantity will depend on the extension of the solution u beyond $\Omega + \mathcal{C}$. In order to avoid problems with this, we will restrict our attention to exterior data with support inside $\Omega_{e,\mathcal{C}}$, so f and then u admit the extension by zero outside $\Omega + \mathcal{C}$.

With all the previous considerations, we define the following Dirichlet-to-Neumann map:

$$\begin{aligned} \Lambda^s_{\mathcal{C},q} : \tilde{H}^s(\Omega_{e,\mathcal{C}}) &\longrightarrow \left(\tilde{H}^s(\Omega_{e,-\mathcal{C}})\right)^*, \\ f &\longmapsto Au|_{\Omega_{e,-\mathcal{C}}}, \end{aligned} \tag{23}$$

where u is the (unique) solution to (22).

Similar arguments can be considered for an operator S. Without further details, we pose the problem

$$\begin{aligned} (S+q)u &= 0 \text{ in } \Omega, \\ u &= f \text{ in } \Omega_{e,\mathcal{C}^2}, \end{aligned} \tag{24}$$

where $\Omega_{c,\mathcal{C}^2} = (\Omega + \mathcal{C}^2)\backslash\bar{\Omega}$, and we define the Dirichlet-to-Neumann map

$$\begin{aligned} \Lambda^s_{\mathcal{C}^2,q} : \tilde{H}^s(\Omega_{e,\mathcal{C}^2}) &\longrightarrow \left(\tilde{H}^s(\Omega_{e,\mathcal{C}^2})\right)^*, \\ f &\longmapsto Su|_{\Omega_{e,\mathcal{C}^2}}, \end{aligned} \tag{25}$$

where u is the (unique) solution to (24).

3.2.1 Runge Approximation Results for Antilocal-In-Cones Operators

Antilocality in cones implies Runge-type approximation results similar to (20) but with additional geometric conditions on the set U:

Theorem 3.1. ([7], Theorem 2 and Theorem 5.25)bf . *Let $\Omega \subset \mathbb{R}^n$ be a bounded, Lipschitz open set and let $U \subset \Omega_{e,\mathcal{C}}$ / $U \subset \Omega_{e,\mathcal{C}^2}$ be a non-empty bounded subset such that $\Omega \subset U - \mathcal{C}$ / $\Omega \subset U + \mathcal{C}^2$. Assume the adjoint operator A^* / S^* is $(-\mathcal{C})$-antilocal*[12] */ $\mathcal{C}^2$-antilocal and $q \in L^\infty(\Omega)$ is such that the problem* (22) */*(24) *and the corresponding problem for the adjoint operator are well-posed. Then for any function $v \in \tilde{H}^s(\Omega)$ there exists a solution of* (22) *with* $\mathrm{supp}(f) \subseteq U$ *such that $v - u|_\Omega$ is arbitrarily small (in the $\tilde{H}^s(\Omega)$ norm).*

---

[11] We assume that for a given $f \in H^s(\Omega_{e,\mathcal{C}})$ there exists a unique (weak) solution $u \in H^s(\Omega + \mathcal{C})$. For A as in (21), well-posedness holds provided 0 does not belong to the (countable) set of eigenvalues [7].

[12] Notice that the operator A^* "sees" the cone $-\mathcal{C}$.

Figure 7 illustrates the geometric conditions on U and Ω. The first one, $U \subset \Omega_{e,\mathcal{C}}$ for A, is natural in view of the Dirichlet problem (22). The second one, $\Omega \subset U - \mathcal{C}$, arises as an additional necessary condition from the $\mathcal{C}$-antilocal property.

If A and S are those of (21), we know that $S^* = S$ while A^* agrees with A but replacing $\mathcal{C}$ by $-\mathcal{C}$. Then, the conditions on antilocality in cones and well-posedness for the adjoints can be directly inferred from those for the operators themselves.

We finally remark that $\mathcal{C}^2$-antilocality implies weak unique continuation for the equation $S + q$ in regular enough domains Ω: If u is a solution of (24) and $u = 0$ in $U \subset \Omega$, then $u = 0$ in $\Omega + \mathcal{C}^2$ ([7, Proposition 4.3]). However, a similar version of unique continuation for a $\mathcal{C}$-antilocal operator may fail ([7, Lemma 5.27]).

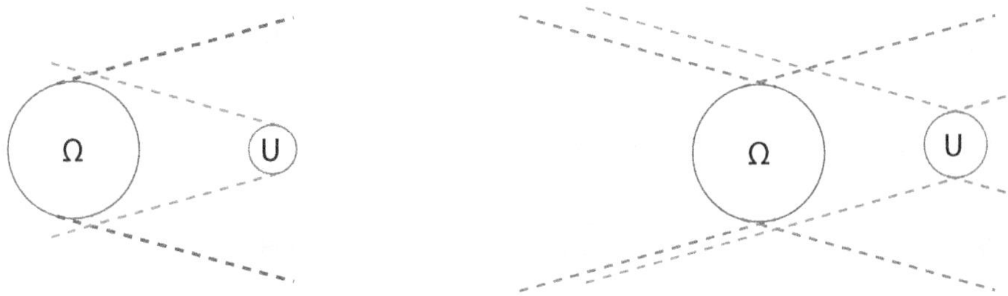

Fig. 7. Example of settings for which the geometric conditions of Theorem 3.1 are satisfied, in the case of A (left) and S (right).

3.2.2 Uniqueness for the Partial-Data Calderón Problem

Arguing as in the fractional Calderón problem and using Runge Theorem 3.1, we obtain uniqueness from partial data (and infinite measurements) of the Calderón problems $\Lambda^s_{\mathcal{C},q} \mapsto q$ and $\Lambda^s_{\mathcal{C}^2,q} \mapsto q$ under some geometric conditions:

Theorem 3.2. ([7], Theorem 3 and Proposition 5.29). *Let Ω, V, W be bounded, Lipschitz open set such that $V \subset \Omega - \mathcal{C}$, $W \subset \Omega + \mathcal{C}$ and $\Omega \subset (V + \mathcal{C}) \cap (W - \mathcal{C})$. Let $q_1, q_2 \in L^\infty(\Omega)$. Assume that A and A^* are $\mathcal{C}$- and $(-\mathcal{C})$-antilocal, respectively, and that the Dirichlet problems for $A + q_i$ and $A^* + q_i$ are well-posed. Then $\Lambda^s_{\mathcal{C},q_1} f|_V = \Lambda^s_{\mathcal{C},q_2} f|_V$ for any $f \in C^\infty_c(W)$ implies that $q_1 = q_2$.*

The same conclusion holds from $\Lambda^s_{\mathcal{C}^2,q_1} f|_V = \Lambda^s_{\mathcal{C}^2,q_2} f|_V$ under analog assumptions on S and the geometric conditions $V, W \subset \Omega + \mathcal{C}^2$ and $\Omega \subset (V + \mathcal{C}^2) \cap (W + \mathcal{C}^2)$.

The geometric conditions on W and V are required in order to apply Theorem 3.1 for $A + q_1$ with $U = W$ and for $A^* + q_2$ with $U = V$. See Fig. 8 for an illustrative example of the geometric setting.

Finally, we state uniqueness from only one single measurement for regular enough potentials:

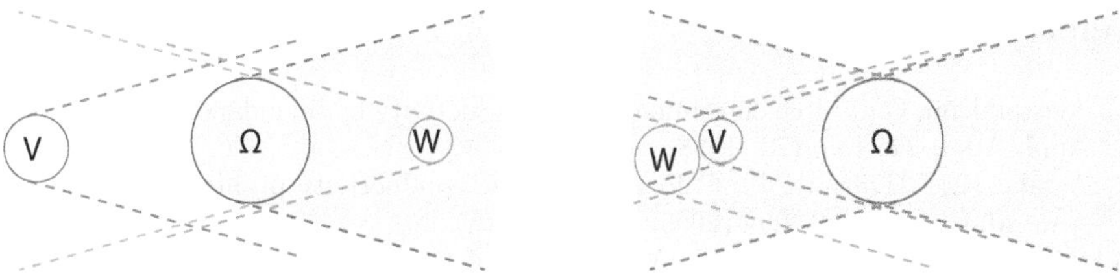

Fig. 8. Example of geometric settings for Theorem 3.2, for A (left) and S (right).

Theorem 3.3. ([7], Theorem 4 and Proposition 5.30). *Let Ω, V, W be bounded, C^1 open set such that $V \subset \Omega - \mathcal{C}$, $W \subset \Omega + \mathcal{C}$, $\Omega \subset V + \mathcal{C}$ and $W \subset \cap_{x \in \Omega}(x + \mathcal{C})$. Let $q \in C^0(\Omega)$. Assume that A is $\mathcal{C}$-antilocal and that the Dirichlet problem* (22) *is well-posed. Let $f \in C_c^\infty(W)$ be a non-trivial function. Then the knowledge of f and $\Lambda^s_{\mathcal{C},q} f|_V$ determines q uniquely.*

The same conclusion holds from $\Lambda^s_{\mathcal{C}^2,q} f|_V$ under analog assumptions on S and the geometric conditions $V, W \subset \Omega + \mathcal{C}^2$ and $\Omega \subset V + \mathcal{C}^2$.

We remark here that the extra condition on W in the case of the operator A is imposed to avoid the (possible) lack of weak unique continuation, which plays a crucial role in the proof. As we said above, $\mathcal{C}^2$-antilocality implies unique continuation provided the domain Ω is regular enough (C^1) and no additional conditions are required. Figure 9 illustrates this particular geometric setting.

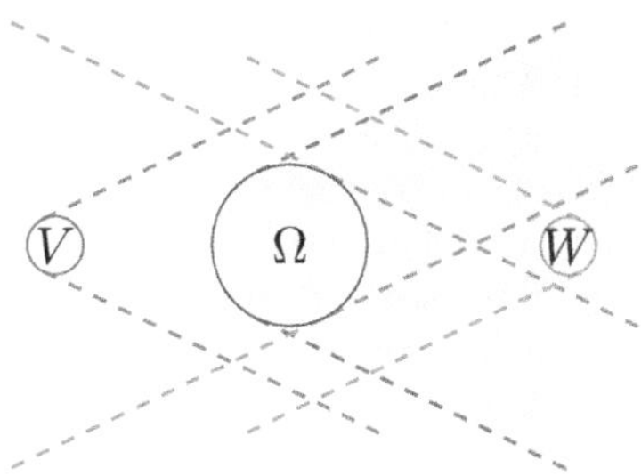

Fig. 9. Example of geometric settings for the first part of Theorem 3.3. Namely, W must lie in the right golden cone.

Acknowledgments. MAGF would like to thank Giovanni Covi, Angkana Rüland and Wiktoria Zatoń, coauthors of the kernel results of this exposition, and the RSME, instigator of collecting in writing the ideas shared during the talk she gave upon their kind invitation. During the writing process of these notes, MAGF has been partially supported by the BBVA Foundation through the Start-up Grant José Luis Rubio de Francia 2019 and by the AEI projects PID2021-125021NAI00, PID2021-124195NB-C32, PID2021-122154NB-I00 and PID2021-122156NB-I00.

References

1. Alessandrini, G.: Stable determination of conductivity by boundary measurements. Appl. Anal. **27**, 153–172 (1988)
2. Astala, K., Päivärinta, L.: Calderón's inverse conductivity problem in the plane. Ann. Math. **163**, 265–299 (2006)
3. Barceló, T., Faraco, D., Ruiz, A.: Stability of Calderón's inverse problem in the plane. J. Math. Pures Appl. **88**, 522–556 (2007)
4. Browder, F.E.: Approximation by solutions of partial differential equations. Amer. J. Math. **84**, 134–160 (1962)
5. Calderón, A.P.: On an inverse boundary value problem, Seminar on Numerical Analysis and its Applications to Continuum Physics, pp. 65–73. Río de Janeiro, Soc. Brasileira de Matemática (1980)
6. Caro, P., Rogers, K.M.: Global uniqueness for the Calderón problem with Lipschitz conductivities. Forum Math. Pi **4**(28), 1–28 (2016)
7. Covi, G., García-Ferrero, M.Á., Rüland, A.: The Calderón problem for nonlocal Schrödinger equations with homogeneous, directionally antilocal principal symbols. J. Differ. Equ. **341**, 79–149 (2022)
8. Feldman, J., Salo, M., Uhlmann, G.: The Calderón Problem - An Introduction to Inverse Problems, partially completed textbook
9. García-Ferrero, M.Á., Rüland, A., Zatoń, W.: Runge approximation and stability improvement for a partial data Calderón problem for the acoustic Helmholtz equation, Inverse Probl. Imaging (2021) 1–31 (2021)
10. Ghosh, T., Rüland, A., Salo, M., Uhlmann, G.: Uniqueness and reconstruction for the fractional Calderón problem with a single measurement. J. Funct. Anal. **279**(1), 108505 (2020)
11. Ghosh, T., Salo, M., Uhlmann, G.: The Calderón problem for the fractional Schrödinger equation. Anal. PDE **13**(2), 455–475 (2020)
12. Haberman, B., Tataru, D.: Uniqueness in Calderón's problem with Lipschitz conductivities. Duke Math. J. **162**(3), 497–516 (2013)
13. Isakov, V.: Increasing stability for the Schrödinger potential from the Dirichlet-to-Neumann map. Discrete Cont. Dyn. Syst. S **4**, 631–640 (2011)
14. Isakov, V., Nagayasu, S., Uhlmann, G., Wang, J.-N.: Increasing stability of the inverse boundary value problem for the Schrödinger equation. Contemp. Math. **615**, 131–141 (2014)
15. Ishikawa, Y.: Antilocality and one-sided antilocality for stable generators on the line. Tsukuba J. Math. **10**(1), 1–9 (1986)
16. Ishikawa, Y.: The Γ-antilocality of stable generators whose Lévy measures are supported on a cone. Tokyo J. Math. **12**(1), 131–143 (1989)
17. Koch, H., Rüland, A., Salo, M.: On instability mechanisms for inverse problems, Ars Inveniendi Analytica (2021) 7
18. Kohn, R., Vogelius, M.: Identification of an unknown conductivity by means of measurements. Inverse Probl. SIAM- AMS Proc. **14**, 113–123 (1984)
19. Krupchyk, K., Uhlmann, G.: Stability estimates for partial data inverse problems for Schrödinger operators in the high frequency limit. J. Math. Anal. Appl. **126**, 273–291 (2019)
20. Lax, P.D.: A stability theorem for solutions of abstract differential equations, and its application to the study of the local behavior of solutions of elliptic equations. Comm. Pure Appl. Math. **9**, 747–766 (1956)

21. Malgrange, B.: Existence et approximation des solutions des équations aux dérivées partielles et des équations de convolution, Ann. Inst. Fourier **6**, 271–355 (1955–1956)
22. Mandache, N.: Exponential instability in an inverse problem for the Schrödinger equation. Inverse Probl. **17**, 1435–1444 (2001)
23. Nachman, A.: Reconstructions from boundary measurements. Ann. Math. **128**, 531–576 (1988)
24. Nagayasu, S., Uhlmann, G., Wang, J.-N.: Increasing stability in an inverse problem for the acoustic equation. Inverse Probl. **29**, 025012 (2013)
25. Rüland, A.: Mikko Salo, The fractional Calderón problem: Low regularity and stability, Nonlinear. Analysis **193**, 111529 (2020)
26. Runge, C.: Zur Theorie der eindeutigen analytischer Funktionen. Acta Math. **6**, 229–244 (1885)
27. Sylvester, J., Uhlmann, G.: A global uniqueness theorem for an inverse boundary value problem. Ann. of Math. **125**(1), 153–169 (1987)
28. Sylvester, J., Uhlmann, G.: Inverse boundary value problems at the boundary-continuous dependence. Comm. Pure Appl. Math. **41**, 197–221 (1998)
29. Uhlmann, G.: 30 Years of Calderón's Problem, Séminaire Laurent Schwartz - EDP et applications (2012-2013), Talk no. 13
30. Uhlmann, G.: Inverse problems: seeing the unseen. Bull. Math. Sci. **4**, 209–279 (2014)
31. Tataru, D.: Unique continuation for PDE's. IMA Vol. Math. Appl. **137**, 239–255 (2003)

Free Objects in Banach Space Theory

Enrique García-Sánchez, David de Hevia, and Pedro Tradacete(✉)

Instituto de Ciencias Matemáticas (CSIC-UAM-UC3M-UCM), Consejo Superior de Investigaciones Científicas, C/ Nicolás Cabrera, 13–15, Campus de Cantoblanco UAM, 28049 Madrid, Spain
{enrique.garcia,david.dehevia,pedro.tradacete}@icmat.es

Abstract. We survey recent progress on three relevant instances of free objects related to Banach spaces: Lipschitz free spaces generated by metric spaces, holomorphic free spaces generated by open sets and free Banach lattices generated by Banach spaces. Our emphasis will be on the parallelisms among these developments.

Keywords: Free object · Lipschitz free space · free Banach lattice

2020 Mathematics Subject Classification. 46B42 · 46B10 · 46G20 · 46E15 · 46B20

1 Introduction

The aim of this note is to review recent developments concerning free objects arising in different aspects of analysis and which are deeply connected to Banach space theory. The first of these notions that we will revisit here are *Lipschitz free Banach spaces*, which associate a Banach space to a metric space. This construction has been the object of intense research in the last 20 years and has been useful in the study of the non-linear geometry of Banach spaces, with ramifications and applications in metric embeddings, geometric group theory... (cf. [26,33,75]). Secondly, we will review what we will call *holomorphic free Banach spaces*, which originate in a construction due to J. Mujica [71], allowing to linearize holomorphic maps from an open set to a Banach space, and which, although not so developed, somehow parallel the theory of Lipschitz free spaces. Finally, we will focus on *free Banach lattices*, which are a recent addition due to A. Avilés, J. Rodríguez and P. Tradacete [21] and provide a new tool for studying the interactions between lattice structures and Banach spaces.

Research partially supported by grants PID2020-116398GB-I00 and CEX2019-000904-S funded by MCIN/AEI/10.13039/501100011033. E. García-Sánchez was partially supported by FPI grant CEX2019-000904-S-21-3 funded by MCIN/AEI/10.13039/501100011033 and by "European Union NextGenerationEU/PRT". D. de Hevia benefited from an FPU Grant FPU20/03334 from the Ministerio de Universidades. P. Tradacete is also supported by a 2022 Leonardo Grant for Researchers and Cultural Creators, BBVA Foundation.

H. Herrero et al. (Eds.): RSME 2022, RSME 13, pp. 100–124, 2024.
https://doi.org/10.1007/978-3-031-62025-6_6

All these examples can be framed within the notion of the *free object* in a certain category generated by an object from a (typically larger) category, which has been central in several developments in different areas of mathematics. Probably, the most significant and better known examples are related to universal algebra, and include in particular free groups, free modules, free algebras or free lattices.

The language of Category theory is particularly convenient to deal with this notion. The idea of a free object can be somehow seen as a generalization of the notion of basis of a vector space, based on the property that every linear map is determined by the values on the basis, and has its origins in the work of S. MacLane [68], J. R. Isbell [55] and Z. Semadeni [83], among others. The extension to the categories setting can be expressed as follows: Suppose we have categories $\mathcal{C}_1$ and $\mathcal{C}_2$, with a faithful functor $f : \mathcal{C}_2 \to \mathcal{C}_1$; typically, we will consider $\mathcal{C}_1$ as a category which is "larger" than $\mathcal{C}_2$ (for instance, we could take $\mathcal{C}_1$ as the category of metric spaces with Lipschitz functions and $\mathcal{C}_2$ that of Banach spaces with bounded linear operators) and consider f as the "forgetful" functor (which in the same example would take a Banach space $(X, \|\cdot\|)$ to the metric space $(X, d(\cdot,\cdot))$, with $d(x, y) = \|x - y\|$, and a bounded linear operator to the corresponding Lipschitz map); now, given an object O in $\mathcal{C}_1$, the *free object in* $\mathcal{C}_2$ *generated by* O, is an object $F(O)$ in $\mathcal{C}_2$ together with a morphism $i : O \to f(F(O))$ satisfying the following universal property: for any object X in $\mathcal{C}_2$ and any morphism $\varphi : O \to f(X)$ in $\mathcal{C}_1$, there is a unique morphism $F(\varphi) : F(O) \to X$ in $\mathcal{C}_2$ such that $f(F(\varphi)) \circ i = \varphi$. It will be customary to plot this in the shape of the following commuting diagram:

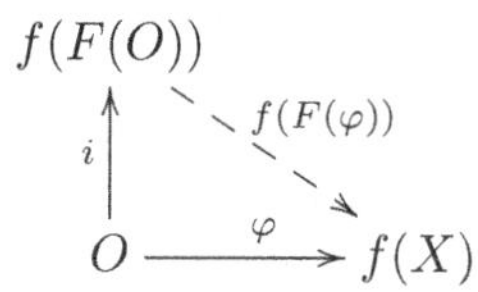

Free objects have also played a fundamental role in topology, although these can be found in the literature under somewhat different terminologies. For instance, recall that the *Stone-Čech compactification* of a topological space S is a compact Hausdorff space βS together with a continuous map $i_S : S \to \beta S$, so that for every compact Hausdorff space K and every continuous map $f : S \to K$ there is a unique continuous map $\beta f : \beta S \to K$ making the following diagram commute:

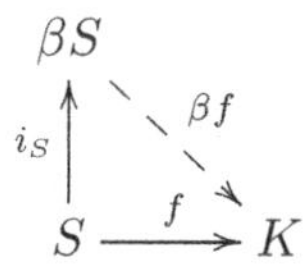

Thus, we could consider βS as the *free compact Hausdorff space generated by* S.

A similar scheme can be used to describe different notions of completions or closures of a given space. These include in particular the completion of a metric, or uniform, space (i.e. adding limits of Cauchy sequences) or the order completion of lattices or boolean algebras (by adding infima and suprema of order bounded sets).

The study of free objects in Banach space theory can be considered as a more recent development. However, we will see next how the second dual of a Banach space can be framed in this picture. Consider the category of *dual* Banach spaces together with *adjoint operators* (or equivalently, weak* continuous linear maps). Given a Banach space E, let $J_E : E \to E^{**}$ denote the canonical embedding; it is clear that every bounded linear operator $T : E \to X^*$ can be uniquely extended to a dual operator $\hat{T} : E^{**} \to X^*$, given by $\hat{T} = (T^* \circ J_X)^*$, in such a way that the following diagram commutes:

$$\begin{array}{ccc} E^{**} & & \\ {\scriptstyle J_E}\uparrow & \searrow^{\hat{T}} & \\ E & \xrightarrow{\;T\;} & X^* \end{array}$$

Thus, we can consider E^{**} (together with the embedding J_E) as the *free dual Banach space generated by* E.

Note this construction however, is actually closer to that of a completion: if E is a Banach space and we consider its unit ball B_E equipped with the weak topology (the one induced by the functionals in E^*), then $B_{E^{**}}$ can be considered as the completion with respect to that topology (which actually makes the ball into a compact set by Banach-Alaoglu's theorem).

The paper is organized as follows. Section 2 is devoted to Lipschitz free Banach spaces generated by a pointed metric space. We will recall first the different ways to construct this space and then focus on the interaction between properties of the Lipschitz free space $\mathcal{F}(M)$ and the underlying metric space M. In Sect. 3 we introduce the notion of holomorphic free Banach space $\mathcal{G}^\infty(U)$ generated by an open subset U of complex Banach space. After showing the construction of this object, we present an overview about its main properties. Finally, Sect. 4 focuses on the free Banach lattice generated by a Banach space, following a similar structure as in the previous sections.

Finally, note that an abstract approach to free objects can be useful for identifying general principles that have different manifestations in each of the aforementioned constructions. This approach has been pursued for instance in [16] exhibiting the relation between local complementation and extension properties in $\mathcal{F}(M)$ and FBL$[E]$. A general construction for linearizing functions with values in locally convex spaces has also been considered in [32] which covers in particular the case of $\mathcal{G}^\infty(U)$.

We refer the reader to the monographs [40, 66] for Banach space theory, [67, 70] for Banach lattices and [26, 86] for the theory of Lipschitz functions.

2 Lipschitz Free Banach Space

In the recent years, especially after the appearance of [86], *Lipschitz free spaces* (also known as *Arens-Eells spaces* or *transportation cost spaces*) have attracted considerable attention from researchers interested in metric geometry and the non-linear theory of Banach spaces (see, for instance, the survey paper [47], and also [77, 1.6] for historical remarks about this construction). These spaces can be defined by the following universal property: Given a metric space M with a distinguished point 0, $\mathcal{F}(M)$ is a Banach space together with an isometric map $\delta : M \to \mathcal{F}(M)$ with the property that for every Banach space X and every Lipschitz map $f : M \to X$ with $f(0) = 0$, there is a unique linear operator $\widehat{f} : \mathcal{F}(M) \to X$ making the following diagram commute:

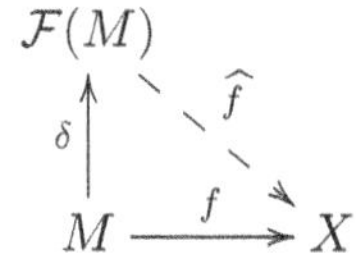

Moreover, the operator norm of the linear extension coincides with the Lipschitz norm of the function f: $\|\widehat{f}\| = \|f\|_{Lip}$. A very fruitful line of research is devoted to analyzing the interplay between Banach space properties of $\mathcal{F}(M)$ versus metric properties of M (see for example [6,8,34,35,46,48,49]).

2.1 Construction

Let (M, d) be a metric space with a distinguished point denoted by 0 and let $\mathrm{Lip}_0(M)$ denote the space of real-valued Lipschitz functions that map $0 \in M$ to $0 \in \mathbb{R}$. We can equip this space with the norm

$$\|f\|_{\mathrm{Lip}} = \sup\left\{\frac{|f(x) - f(y)|}{d(x, y)} : x \neq y \in M\right\},$$

for $f \in \mathrm{Lip}_0(M)$, so that $\mathrm{Lip}_0(M)$ becomes a Banach space.

For $x \in M$, let $\delta_x \in \mathrm{Lip}_0(M)^*$ be the evaluation functional given by $\delta_x(f) = f(x)$ for $f \in \mathrm{Lip}_0(M)$. Let us denote by $\mathcal{F}(M)$ the closed linear span of $\{\delta_x : x \in M\}$ with the dual norm in $\mathrm{Lip}_0(M)^*$. Note that the map $x \mapsto \delta_x$ becomes an isometry, as we have $\|\delta_x - \delta_y\| = d(x, y)$ for any $x, y \in M$.

The space $\mathcal{F}(M)$ together with $\delta_M : M \to \mathcal{F}(M)$ is the *Lipschitz free Banach space* and it is uniquely characterized by the following universal property:

For every Banach space X and every Lipschitz map $f : M \to X$ with $f(0) = 0$, there is a unique linear operator $\widehat{f} : \mathcal{F}(M) \to X$ making the following diagram commute:

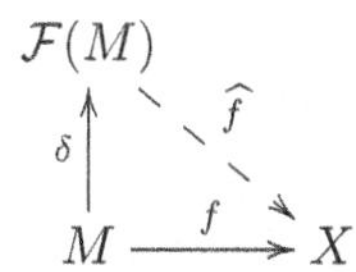

Let us briefly recall the proof of this fact. Fix a Banach space X and a Lipschitz map $f : M \to X$ mapping 0 to 0. Extend linearly f to $\mathrm{span}\{\delta_x : x \in M\}$ and denote this extension by $\widehat{f}$. It is enough to check that $\|\widehat{f}\| = \|f\|_{\mathrm{Lip}}$. To this end, pick $a \in \operatorname{span}\{\delta_x : x \in M\}$. Then $\|\widehat{f}(a)\|_X = x^*(\widehat{f}(a))$ for some $x^* \in B_{X^*}$. However, $x^* \circ f$ belongs to $\operatorname{Lip}_0(M)$ and $\|x^* \circ f\|_{\mathrm{Lip}} \leq \|f\|_{\mathrm{Lip}}$, so $\|a\|\|f\|_{\mathrm{Lip}} \geq \|\widehat{f}(a)\|_X$ as claimed. Hence $\widehat{f}$ can be uniquely extended to $\mathcal{F}(M)$. As a consequence we can identify $\mathcal{F}(M)^* = \operatorname{Lip}_0(M)$.

The above universal property can also be extended in a natural way to another one with a more functorial character. Suppose (M, d), (N, d') are pointed metric spaces, then every Lipschitz map $f : M \to N$ mapping 0 to 0 admits a linear extension $\overline{f}$ making the following diagram commutative:

$$\begin{array}{ccc} \mathcal{F}(M) & \overset{\overline{f}}{\dashrightarrow} & \mathcal{F}(N) \\ {\scriptstyle \delta_M}\uparrow & & \uparrow{\scriptstyle \delta_N} \\ M & \xrightarrow{\quad f \quad} & N \end{array}$$

It is not hard to see that the map f is bi-Lipschitz if and only if $\overline{f}$ is an isomorphic embedding, that f is a Lipschitz retract if and only if $\overline{f}$ is a linear projection, or that f has dense range if and only if $\overline{f}$ has. The question about injectivity is less clear and has been pursued in [43].

In connection with this, an important consequence of McShane's extension theorem for Lipschitz functions [86, Theorem 1.5.6] is the following: Suppose N is a subspace of a metric space (M, d), let $\iota : N \hookrightarrow M$ denote the formal inclusion, which is clearly Lipschitz, then $\overline{\iota}$ defines a linear isometric embedding of $\mathcal{F}(N)$ as a subspace of $\mathcal{F}(M)$. As a consequence, we get that if N is bi-Lipschitz equivalent to a subset of M, then $\mathcal{F}(N)$ is linearly isomorphic to a subspace of $\mathcal{F}(M)$.

There is another intrinsic way to define Lipschitz free spaces by means of the notion of the Arens-Eells norm or transportation cost spaces, which are closely related to Wasserstein-1 spaces [77]. For further details and motivation we refer to [86, Section 2.3], but we include here the main steps in this idea: For $a \in \operatorname{span}\{\delta_x : x \in M\}$, let us consider

$$|||a||| = \inf\Big\{ \sum_{j=1}^{n} |\alpha_j| d(y_j, z_j) : a = \sum_{j=1}^{n} \alpha_j (\delta_{y_j} - \delta_{z_j}) \Big\}.$$

On the one hand, it can be easily checked that $|||\cdot|||$ is the largest seminorm ρ on $\operatorname{span}\{\delta_x : x \in M\}$ satisfying $\rho(\delta_x - \delta_y) \leq d(x, y)$ for every $x, y \in M$. As the standard norm $\|\cdot\|_{\mathcal{F}(M)}$ on $\mathcal{F}(M)$ satisfies this condition, it follows that $\|\cdot\|_{\mathcal{F}(M)} \leq |||\cdot|||$. In particular, $|||\cdot|||$ is actually a norm and $|||\delta_x - \delta_y||| = d(x, y)$ for $x, y \in M$.

On the other hand, if we consider now the Banach space

$$X = \overline{\operatorname{span}\{\delta_x : x \in M\}}^{|||\cdot|||}$$

and take the map $f : M \to X$ given by $f(x) = \delta_x$, then f is an isometric embedding, and by the universal property extends to a linear map $\widehat{f} : \mathcal{F}(M) \to X$ which is 1-Lipschitz. Hence, it follows that $X = \mathcal{F}(M)$ with $|||\cdot||| = \|\cdot\|_{\mathcal{F}(M)}$, which is the claim.

2.2 Banach Space Properties of $\mathcal{F}(M)$

As a predual of the space of Lipchitz functions, $\mathcal{F}(M)$ has a certain ℓ_1 behavior. It is well known that if $M \subset \mathbb{R}$ is a set of measure zero or if M is a separable ultrametric space, then $\mathcal{F}(M)$ is isomorphic to ℓ_1, see [34, 46].

In [35] the authors showed that for every infinite metric space M, ℓ_1 always embeds as a complemented subspace of $\mathcal{F}(M)$. As a consequence, we get that $\mathcal{F}(M)$ is never isomorphic to a complemented subspace of a $C(K)$ space, or that $\mathcal{F}(M)^*$ is not weakly sequentially complete (see definition below). Further results about the relation with ℓ_1 can be found in [11, 78].

Recall that a Banach space X has the *approximation property* (AP) if the identity id_X of X is in the closure of the finite rank operators on X for the topology of uniform convergence on compact sets. The λ*-bounded approximation property* (λ-BAP) means that there are finite rank operators with norm less than λ that can be chosen in the approximation, and the (1-BAP) is called the *metric approximation property* (MAP).

In [49] it was shown that for a separable metric space M and $\lambda \geq 1$, the Lipschitz free space $\mathcal{F}(M)$ has the λ-BAP if and only if for any Banach space Y and any $f \in \text{Lip}(M, Y^{**})$, there is a net $(f_\alpha) \subset \text{Lip}(M, Y)$ which converges to f in the pointwise-weak* topology with $\|f_\alpha\| \leq \lambda \|f\|$. Note that in the particular case when M is a Banach space, [48, Theorem 5.3] already shows that $\mathcal{F}(X)$ has the λ-BAP if and only if X has the λ-BAP. More recently, it was shown in [65] that for every doubling metric space M, the space $\mathcal{F}(M)$ has BAP, and actually $\mathcal{F}(M)$ has MAP for every compact convex $M \subset \mathbb{R}^n$ (see [80]). Going even one step further, it was shown in [53] that $\mathcal{F}(\mathbb{R}^n)$ and $\mathcal{F}(\ell_1)$ actually have a Schauder basis. On the opposite direction, in [52] an example is given of a compact metric space homeomorphic to the Cantor set whose Lipschitz free space fails the AP.

As noted above, $\mathcal{F}(M)^* = \text{Lip}_0(M)$. It was asked in [86] whether the space $\text{Lip}_0(M)$ has a unique predual. This has been settled affirmatively in [87] for the cases when M has finite diameter or when M is a Banach space.

In connection with duality, it seems to be a difficult problem to identify the extreme points of the unit ball of $\mathcal{F}(M)$. Natural candidates for such points are the elements of the form

$$m_{x,y} = \frac{\delta_x - \delta_y}{d(x,y)},$$

usually called *(elementary) molecules*. It was proved in [44] that the sets of preserved extreme points and denting points of the unit ball of $\mathcal{F}(M)$ coincide. Also, in [45] the authors show that a molecule $m_{x,y}$ is a strongly exposed point of the unit ball of $\mathcal{F}(M)$ if and only if

$$d(x,z) + d(y,z) - d(x,y) \geq C \min\{d(x,z), d(y,z)\}$$

for some $C > 0$ and every $z \in M$. Here, recall that a norm one element x of a Banach space X is called a *strongly exposed point of* B_X if there is $x^* \in X^*$ with $\|x^*\| = 1$ such that every sequence $(x_n) \subset B_X$ with $x^*(x_n) \to x^*(x)$ must be norm convergent to x. More recently, it has been shown in [5] that if M is a proper metric space (where bounded closed subsets are compact), the extreme points are precisely the molecules $m_{x,y}$ satisfying $d(x,z) + d(y,z) > d(x,y)$ for every $z \in M \setminus \{x,y\}$. It follows that all extreme points are preserved extreme points and also exposed points. See also [7,9] for further related results on extreme points. Apparently, it is still unknown whether all extreme points of the unit ball of $\mathcal{F}(M)$ are molecules for an arbitrary metric space M.

A classical geometrical property of a Banach space is the *Daugavet property*, which can be defined by the equation $\|T+I\| = 1+\|T\|$ holding for every rank-one operator T. The Daugavet property of the space $\mathcal{F}(M)$ has been characterized in [45] precisely as the property of the metric space M to be a length space; this complements earlier results from [56]. Recall that a metric space M is called a *length space* if, for each pair of points $x, y \in M$, $d(x,y)$ coincides with the infimum of the length of rectifiable curves joining x and y (when this infimum is attained we say that M is a *geodesic space*).

A Banach space has the *Schur property* when every weakly convergent sequence must be norm-convergent. It was shown in [52] that the free space over a countable compact metric space has the Schur property. In connection with this, a more general sufficient condition for $\mathcal{F}(M)$ to have the Schur property has been provided in [81] strongly based on the work of N. Kalton [61]: Let

$$\operatorname{lip}_0(M) = \left\{ f \in \operatorname{Lip}_0(M) : \lim_{\varepsilon \to 0} \sup_{0<d(x,y)<\varepsilon} \frac{|f(x) - f(y)|}{d(x,y)} = 0 \right\};$$

if $\operatorname{lip}_0(M)$ is 1-norming for $\mathcal{F}(M)$, then $\mathcal{F}(M)$ has the Schur property. More recently, in [6], the Schur property of $\mathcal{F}(M)$ has been characterized by the fact that (the completion of) M is purely 1-unrectifiable, and this is also equivalent to the Radon-Nikodým property of $\mathcal{F}(M)$.

Recall that a Banach space X is called *weakly sequentially complete* if every weakly Cauchy sequence in X is weakly convergent. Also, a Banach space X is *super-reflexive* if no non-reflexive Banach space is finitely representable in X. If M is a compact subset of a superreflexive Banach space, then the space $\mathcal{F}(M)$ is weakly sequentially complete [64]. Also, $\mathcal{F}(X)$ is weakly sequentially complete whenever X is superreflexive [10].

It is clear that if M_1 and M_2 are Lipschitz isomorphic metric spaces, then the corresponding spaces $\mathcal{F}(M_1)$ and $\mathcal{F}(M_2)$ are linearly isomorphic. The converse of this is false as shown in [39], where the authors showed that if K is any infinite metric compact space, then $\mathcal{F}(C(K))$ and $\mathcal{F}(c_0)$ are linearly isomorphic.

The case when the underlying metric space M is already a Banach space is particularly relevant. If X is a Banach space, by construction there is a unique linear map $\beta_X : \mathcal{F}(X) \to X$ (usually called *barycenter map*) extending the identity on X, that is $\beta_X \circ \delta_X = id_X$. It is natural to wonder under which conditions β_X has a linear right inverse. This is equivalent to asking whether the Banach space X linearly embeds as a subspace of $\mathcal{F}(X)$ so that β_X defines a projection onto this subspace. This property was considered in the seminal paper by G. Godefroy and N. Kalton [48], where it was shown that every separable Banach space and every space of the form $\mathcal{F}(M)$ have this property.

The work [48] also provides an important application of Lipschitz free spaces to the non-linear theory of Banach spaces: Namely, if a separable Banach space X isometrically embeds into another Banach space Y, then there exists an isometric linear embedding, while this does not hold in the non-separable setting.

Another interesting feature of $\mathcal{F}(X)$ for X a Banach space proved in [62] is that $\mathcal{F}(X)$ is linearly isomorphic to the infinite ℓ_1-sum of $\mathcal{F}(X)$. As a consequence, it is observed also in [62] that $\mathcal{F}(X)$ is linearly isomorphic to $\mathcal{F}(B_X)$.

Finally, let us note that a quasi-normed version of Lipschitz free spaces has been considered in [4]. Recall, (X, ρ) is a *quasi-metric space* if $\rho : X \times X \to [0, \infty)$ is a symmetric function such that $\rho(x, y) = 0$ if and only if $x = y$ and there exists $k \geq 1$ such that $\rho(x, y) \leq k(\rho(x, z) + \rho(y, z))$ for $x, y, z \in X$. In this direction, a vector space X, together with a map $\|\cdot\| : X \to [0, \infty)$ such that $\rho(x, y) = \|x - y\|$ is a quasi-metric and $\|\lambda x\| = |\lambda|\|x\|$ is called a *quasi-normed space*. The Aocki-Rolewicz theorem states that every quasi-Banach space admits an equivalent p-norm $|||\cdot|||$ (that satisfies $|||x + y|||^p \leq |||x|||^p + |||y|||^p$). Now, for a pointed p-metric space (X, ρ), one can construct the Lipschitz free p-space over (X, ρ) in a similar spirit as that of $\mathcal{F}(M)$ so that the latter becomes the case $p = 1$. This has been further explored in [1–3]. Along similar lines free spaces over assymetric spaces (those where the "distance" function is not necessarily symmetric) have been considered in [36].

3 Holomorphic Free Banach Space

In the same spirit as Lipschitz free Banach spaces, we can construct what could be called *holomorphic free Banach spaces.* This object is a complex Banach space which allows us to extend bounded holomorphic functions defined on a fixed open subset of a (complex) Banach space to continuous $\mathbb{C}$-linear operators. Given complex Banach spaces E and F and an open subset U of E, we say that a function $f : U \to F$ is *holomorphic* if it is Fréchet differentiable at every $x \in U$, that is, for every $x \in U$ there exists a continuous $\mathbb{C}$-linear mapping $Df(x) : E \to F$ such that

$$\lim_{h \to 0} \frac{f(x + h) - f(x) - Df(x)(h)}{\|h\|} = 0. \tag{3.1}$$

For simplicity, throughout this section Banach spaces will always be defined over the field of complex numbers and operators on them will be $\mathbb{C}$-linear.

3.1 Construction

Given an open subset U of a Banach space E, there is a Banach space $\mathcal{G}^\infty(U)$ together with a bounded holomorphic function $\delta_U : U \to \mathcal{G}^\infty(U)$ with the following universal property: for every Banach space F and every bounded holomorphic function $f : U \to F$, there exists a unique operator $\hat{f} : \mathcal{G}^\infty(U) \to F$ such that

$\hat{f} \circ \delta_U = f$ and $\|\hat{f}\| = \|f\|_\infty$, where $\|f\|_\infty = \sup_{x \in U} |f(x)|$. In other words, $\hat{f}$ makes commutative the following diagram:

$$\begin{array}{ccc} \mathcal{G}^\infty(U) & & \\ \delta_U \uparrow & \searrow^{\hat{f}} & \\ U & \xrightarrow{\;f\;} & F \end{array}$$

The existence of such an object was proved by Mujica in [71, Theorem 2.1], inspired by the Dixmier-Ng theorem (see also [38, p. 417]). The uniqueness of $\mathcal{G}^\infty(U)$ (up to an isometric isomorphism) follows directly from its definition. In [71], the author showed that

$$\mathcal{G}^\infty(U) = \{u \in \mathcal{H}^\infty(U)^* \ : \ u|_{(B_{\mathcal{H}^\infty(U)}, \tau_c)} \text{ is continuous}\},$$

where

$$\mathcal{H}^\infty(U) = \{f : U \to \mathbb{C} : f \text{ is holomorphic and bounded}\},$$

τ_c represents the compact-open topology, and

$$\begin{aligned} \delta_U \ : U &\longrightarrow \mathcal{G}^\infty(U) \\ x &\longmapsto \delta_x \end{aligned}$$

is given by $\delta_x(f) = f(x)$, $f \in \mathcal{H}^\infty(U)$.

Remark 3.1. Note that from the universal property (setting $F = \mathbb{C}$) it follows that the holomorphic free Banach space $\mathcal{G}^\infty(U)$ is an isometric predual of $\mathcal{H}^\infty(U)$. A reasonable question that arises is whether for every open subset U of a Banach space E the space $\mathcal{H}^\infty(U)$ has a unique isometric predual (compare to the results mentioned above in Sect. 2 about the predual of $\mathrm{Lip}_0(M)$). T. Ando [12] and P. Wojtaszczyk [88] independently proved that $\mathcal{H}^\infty(\mathbb{D})$ has a unique predual. This has been generalized for other planar domains in [82]. However, it is not known whether $\mathcal{H}^\infty(\mathbb{D}^n)$ has a unique predual for $n > 2$.

A remarkable property of these spaces is that they always satisfy a sort of *lifting property*. Given an open bounded subset U of a Banach space E, we can consider the restriction to U of the identity mapping on E, denoted by $\iota : U \to E$, which is a bounded holomorphic function. By the universal property of $\mathcal{G}^\infty(U)$, there exists an operator β which makes commutative the diagram below

$$\begin{array}{ccc} \mathcal{G}^\infty(U) & & \\ \delta_U \uparrow & \searrow^{\beta} & \\ U & \xrightarrow{\;\iota\;} & E \end{array}$$

It can be shown that there exists a linear operator $\alpha : E \to \mathcal{G}^\infty(U)$ such that $\beta \circ \alpha = id_E$ (cf. [71, Proposition 2.3]). This clearly parallels the results of [48] mentioned in Sect. 2. Consequently, E is isomorphic to a complemented subspace of $\mathcal{G}^\infty(U)$ whenever U is an open bounded set in E. More specifically, we can take $\alpha = D\delta_U(x_0)$ for any $x_0 \in U$, where $D\delta_U(x_0)$ denotes the Fréchet derivative of δ_U at x_0, which is an operator from E into $\mathcal{G}^\infty(U)$ (recall Eq. (3.1)). When $U = B_E$, we can achieve an isometric version of this *lifting property* by choosing $\alpha = D\delta_U(0)$.

In [71, Corollary 4.12], J. Mujica pointed out an alternative description of holomorphic free Banach spaces which is analogous to the identification of Lipschitz free spaces as *transportation cost spaces* mentioned in the previous section. Given an open subset U of a Banach space E, its corresponding holomorphic free Banach space $\mathcal{G}^\infty(U)$ consists of all linear functionals $u \in \mathcal{H}^\infty(U)$ of the form

$$u = \sum_{k=1}^{\infty} \alpha_k \delta_{x_k}, \tag{3.2}$$

where (x_k) is a sequence of elements of U and (α_k) is an absolutely summable sequence of complex numbers. Furthermore, the norm of an element $u \in \mathcal{G}^\infty(U)$ is given by

$$\|u\| = \inf \sum_{k=1}^{\infty} |\alpha_k|,$$

where the above infimum is taken over all possible representations of u in the form (3.2).

In [73, Theorem 6.1], the author inferred from the simple observation that $\mathcal{H}^\infty(U \times V)$ is isometrically isomorphic to $\mathcal{H}^\infty(U, \mathcal{H}^\infty(V))$ and the universal property of projective tensor products (cf. [40, Proposition 16.13]) that $\mathcal{G}^\infty(U \times V) = \mathcal{G}^\infty(U) \otimes_\pi \mathcal{G}^\infty(V)$.

Recall that a sequence $(x_k)_{k=1}^\infty$ of elements of an open subset U of a Banach space E is said to be an *interpolating sequence* if the mapping

$$f \in \mathcal{H}^\infty(U) \mapsto (f(x_k))_{k=1}^\infty \in \ell_\infty$$

is surjective. In [72, Theorem 5.2], J. Mujica relates this notion to holomorphic free Banach spaces, showing that a sequence $(x_k) \subset U$ is interpolating if and only if there exists an operator $T : \mathcal{G}^\infty(U) \to \ell_1$ such that $T \circ S\xi = \xi$ for every $\xi = (\xi_k) \in \ell_1$, where S is the operator defined by

$$\begin{aligned} S : \ell_1 &\longrightarrow \mathcal{G}^\infty(U) \\ (\xi_k) &\longmapsto \textstyle\sum_{k=1}^\infty \xi_k \delta_{x_k}. \end{aligned}$$

In particular, it follows that if U has an interpolating sequence, then $\mathcal{G}^\infty(U)$ has a complemented copy of ℓ_1. This should be compared with the fact that for infinite metric spaces $\mathcal{F}(M)$ always contains a complemented copy of ℓ_1 (see Sect. 2).

3.2 Connection Between Properties of U and $\mathcal{G}^\infty(U)$

We will devote this part of the section to explaining some already known connections between properties of a Banach space E and those corresponding to $\mathcal{G}^\infty(U)$ for U open in E.

The first observation is that the density character of $\mathcal{G}^\infty(U)$ is at most that of the Banach space E containing U. By *density character* of a topological space we mean the smallest cardinal for a dense subset of the space. The above claim can be deduced from the fact that $\delta_U(U)$ generates a dense subspace of $\mathcal{G}^\infty(U)$ (cf. [71, Remark 2.2]). Note also that for any complex Banach space E, by Liouville's theorem $\mathcal{H}^\infty(E) = \mathbb{C}$, so $\mathcal{G}^\infty(E) = \mathbb{C}$ independently of the density character of E.

Approximation properties (see Subsect. 2.2) constitute one of the most studied topics about these connections. Given a balanced, bounded and open subset U in E, Mujica showed that E has the AP if and only if $\mathcal{G}^\infty(U)$ has the AP [71, Theorem 5.4]. In that article, it was pointed out that the aforementioned fact may be helpful for the question of whether $\mathcal{H}^\infty(\mathbb{D})$ has the AP (this problem is formulated in [66, Problem 1.e.10]).

In [71, Proposition 5.7], it is established that a Banach space E has the MAP if and only if $\mathcal{G}^\infty(B_E)$ has the MAP. Despite there is some work addressing the analogous question for BAP [31], it is not known whether E has the BAP if and only if $\mathcal{G}^\infty(B_E)$ has the BAP. We thank Verónica Dimant (Buenos Aires) for clarifying this.

In line with the preceding results, E. Çalişkan also showed that when U is a bounded balanced open subset of a Banach space E, $\mathcal{G}^\infty(U)$ has the CAP if and only if E has the CAP [29]. Where a Banach space E is said to have the *compact approximation property* (CAP for short) if for every $\varepsilon > 0$ and every K compact set in E there is a compact operator $T : X \to X$ such that $\|Tx - x\| \leq \varepsilon$ for all $x \in K$.

Another result worth mentioning is that for any Banach space E the space $\mathcal{G}^\infty(B_E)$ has the Daugavet property, as M. Jung proved in [60, Corollary 2.3] that $\mathcal{H}^\infty(B_E)$ has the Daugavet property and this is inherited by the predual. In particular, $\mathcal{G}^\infty(B_E)$ fails to have the Radon-Nikodým property. As $\mathcal{H}^\infty(B_E)$ cannot have the Radon-Nikodým property either, its predual $\mathcal{G}^\infty(B_E)$ is not an Asplund space (this was previously mentioned in [42, Example 3.5]). Nevertheless, we still know very little about general properties of $\mathcal{G}^\infty(U)$.

3.3 Connection Between Properties of a Holomorphic Function and Its Linearization

Another issue which must be addressed when we are dealing with free objects is how certain properties of bounded holomorphic functions are *transferred* to their corresponding linear extensions (defined on $\mathcal{G}^\infty(U)$) and vice versa. Let us highlight some examples that illustrate this situation:

- Let E and F be Banach spaces and let U be an open subset of E. A bounded holomorphic mapping $f : U \to F$ has finite rank if and only if its corresponding operator $\hat{f} : \mathcal{G}^\infty(U) \to F$ has finite rank [71, Proposition 3.1]. By a *finite rank* holomorphic mapping f we mean a function such that $\text{span}(f(U)) \subset F$ is finite dimensional.

- A bounded holomorphic function $f : U \to F$ has a relatively compact range (resp. relatively weakly compact range) if and only if the corresponding operator $\hat{f} : \mathcal{G}^\infty(U) \to F$ is compact (resp. weakly compact) [71, Proposition 3.4]. More results in this direction can be found in [58].
- A bounded holomorphic function $f : U \to F$ is *Cohen strongly p-summing holomorphic* if and only if its corresponding linearization $\hat{f} : \mathcal{G}^\infty(U) \to F$ is *strongly p-summing* (see [59]).

Consider now $U \subset E$ and $V \subset F$ open subsets of the Banach spaces E and F, respectively. Given a holomorphic function $f : U \to V$, we may consider $\delta_V \circ f : U \to \mathcal{G}^\infty(V)$, which is a holomorphic function (as it is a composition of two holomorphic functions). Then, we can define $\overline{f} : \mathcal{G}^\infty(U) \to \mathcal{G}^\infty(V)$ as the unique linear operator which makes commutative the following diagram:

$$\begin{array}{ccc} \mathcal{G}^\infty(U) & \overset{\overline{f}}{\dashrightarrow} & \mathcal{G}^\infty(V) \\ {\scriptstyle \delta_U}\uparrow & & \uparrow{\scriptstyle \delta_V} \\ U & \xrightarrow{\quad f \quad} & V \end{array}$$

As we have done in the case of Lipschitz free Banach spaces, it is natural to wonder about the relations between f and $\overline{f}$. In [73, Theorem 5.2] J. Mujica showed that the *surjectivity* of a holomorphic mapping $f : U \to V$ implies the surjectivity of the operator $\overline{f} : \mathcal{G}^\infty(U) \to \mathcal{G}^\infty(V)$. In contrast, we cannot make an analogous claim about the *injectivity*. For example, let $i : \mathbb{D} \hookrightarrow \mathbb{C}$ be the canonical inclusion of the unit disc $\mathbb{D}$ into the complex plane $\mathbb{C}$. Then $\overline{i} : \mathcal{G}^\infty(\mathbb{D}) \to \mathcal{G}^\infty(\mathbb{C})$ is not injective, since $\mathcal{G}^\infty(\mathbb{C})$ is one-dimensional (as the only elements of $\mathcal{H}^\infty(\mathbb{C})$ are constant functions by Liouville's theorem).

Moreover, it can be readily checked that f is *biholomorphic* if and only if $\overline{f}$ is a *surjective linear isometry*. Hence, a natural question motivated by the Lipschitz free spaces case is whether $\mathcal{G}^\infty(U)$ and $\mathcal{G}^\infty(V)$ being isometrically isomorphic implies that their corresponding open sets U and V are biholomorphic. As we shall see below, this is not true in general.

Example 3.2. Let us denote by $\mathbb{D}^2 = \{(z_1, z_2) \in \mathbb{C}^2 : |z_1|, |z_2| < 1\}$ the complex bidisc. We define $U = 3/2\,\mathbb{D}^2$ and $V = U \backslash \{(0,0)\}$. It is clear that the open sets U and V are not biholomorphic (they are not even homeomorphic as topological spaces).

By Hartogs's extension theorem, given $f \in \mathcal{H}^\infty(V)$, there exists $F \in \mathcal{H}^\infty(U)$ (which is uniquely determined by the Identity Principle) such that $F|_V = f$. Thus, we can define an operator $T : \mathcal{H}^\infty(V) \to \mathcal{H}^\infty(U)$ by $Tf = F$. Its linearity can be deduced from the Identity Principle and it is also clear that T is norm preserving and surjective.

Finally, we are going to show that T is weak*-to-weak* continuous. By Banach-Dieudonné theorem it is enough to prove that the restriction $T|_{B_{\mathcal{H}^\infty(V)}}$ is weak*-to-weak* continuous. Let $\{f_\alpha\}$ be a weak*-convergent net to some f in $B_{\mathcal{H}^\infty(V)}$. We have to show that $\{Tf_\alpha\}$ converges to Tf in $(\mathcal{G}^\infty(U), w^*)$. Since

$\{Tf_\alpha\}$ is bounded and $\text{span}\{\delta_U(x) : x \in U\}$ is dense in $\mathcal{G}^\infty(U)$, we only need to prove that

$$F_\alpha(x) = \delta_U(x)(Tf_\alpha) \to \delta_U(x)(Tf) = F(x), \qquad \text{for every } x \in U.$$

Since T extends functions from V to U, the only remaining obstacle is satisfying $F_\alpha(0) \to F(0)$. This can be done by using Cauchy's integral formula. As F is holomorphic on U and $\overline{\mathbb{D}} \subset U$, we have that

$$F(0,0) = \frac{1}{(2\pi i)^2}\int_{\partial\mathbb{D}}\int_{\partial\mathbb{D}} \frac{F(w_1,w_2)}{w_1 w_2} dw_1 dw_2 = \frac{1}{(2\pi i)^2}\int_{\partial\mathbb{D}}\int_{\partial\mathbb{D}} \frac{f(w_1,w_2)}{w_1 w_2} dw_1 dw_2,$$

and hence

$$|F_\alpha(0,0) - F(0,0)| \leq \|f_\alpha - f\|_{\partial\mathbb{D}\times\partial\mathbb{D},\infty} \to 0.$$

Remark 3.3. It seems to be an open question whether Banach spaces E and F with holomorphic free Banach spaces $\mathcal{G}^\infty(B_E)$, $\mathcal{G}^\infty(B_F)$ being isometrically isomorphic implies that the unit balls B_E, B_F are biholomorphically equivalent. Since W. Kaup and H. Upmeier [63] (see also [13]) showed that two complex Banach spaces are isometrically isomorphic if and only if their open unit balls are biholomorphically equivalent, the preceding question could be rephrased as follows: are there non-isometrically isomorphic Banach spaces E, F such that $\mathcal{G}^\infty(B_E)$, $\mathcal{G}^\infty(B_F)$ are isometric?

Finally, we would like to point out some other linearization theorems for certain classes of holomorphic functions which have been considered in the literature. For instance, in [69] (see also [74]) it is shown that for every open subset U of a locally convex space E, there exist a complete locally convex space $\mathcal{G}(U)$ and a holomorphic mapping $\delta_U : U \to \mathcal{G}(U)$ such that for each complete locally convex space F and each holomorphic function $f : U \to F$, there is a unique linear mapping $\hat{f} : \mathcal{G}(U) \to F$ such that $\hat{f} \circ \delta_U = f$. Approximation properties of these spaces have been investigated in [28, 30]. There exists an analogous construction for the more restrictive class of *holomorphic functions of bounded type* which can be found in [41, 73]. A similar study has also been carried out for *weighted holomorphic functions* in [50, 51]. More recently, a *free holomorphic Lipschitz space* has been constructed in [14]. This allows the extension of holomorphic Lipschitz maps to linear operators combining somehow the approaches in this and the previous sections.

4 Free Banach Lattice

Let us consider now the category of Banach lattices with lattice homomorphisms and the larger category of Banach spaces with bounded linear operators. Given a Banach space E, we can define the *free Banach lattice generated by E*. This will be a Banach lattice $\mathrm{FBL}[E]$ together with a linear isometric embedding $\delta_E : E \to \mathrm{FBL}[E]$, such that for every Banach lattice X and every linear and bounded operator $T : E \to X$ there is a unique lattice homomorphism $\hat{T}$ making the following diagram commutative:

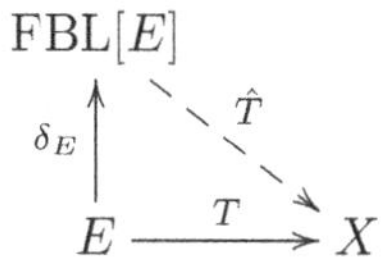

Moreover, $\|\hat{T}\| = \|T\|$.

4.1 Construction

There exists an explicit construction of FBL$[E]$ provided by A. Avilés, J. Rodríguez and P. Tradacete in [21]. Namely, FBL$[E]$ can be represented as a sublattice of the space of positively homogeneous functions from the dual space E^* to $\mathbb{R}$, denoted by $H[E]$, which is a vector lattice when endowed with the pointwise order and lattice operations. It suffices to consider the functional

$$\text{(4.1)}\quad \|f\|_{\mathrm{FBL}[E]} := \sup\left\{\sum_{i=1}^{m} |f(x_i^*)| : m \in \mathbb{N}, (x_i^*)_{i=1}^m \subset E^*, \sup_{x\in B_E}\sum_{i=1}^{m} |x_i^*(x)| \le 1\right\}$$

and define the set $H_1[E] := \left\{f \in H[E] : \|f\|_{\mathrm{FBL}[E]} < \infty\right\}$, which is a Banach lattice with the norm $\|\cdot\|_{\mathrm{FBL}[E]}$. Note that the evaluation functions $\delta_x : E^* \to \mathbb{R}$ with $x \in E$, given by $\delta_x(x^*) = x^*(x)$ for $x^* \in E^*$, define a linear isometry

$$\begin{aligned} \delta_E : E &\longrightarrow H_1[E] \\ x &\longmapsto \delta_E(x) = \delta_x \end{aligned}$$

It can be checked that the (not necessarily closed) sublattice generated by $\delta_E(E)$ in $H[E]$, denoted by FVL$[E]$, is the *free vector lattice over E (as a vector space)*, or equivalently, it satisfies the following universal property: given any vector lattice X and any linear operator $T : E \to X$, there exists a unique lattice homomorphism $\hat{T} : \mathrm{FVL}[E] \to X$ such that $\hat{T} \circ \delta_E = T$. Note that free vector lattices were already considered in [25,27] and their construction follows a standard process. That this vector lattice actually support a lattice norm is less evident.

More precisely, the closure of FVL$[E]$ under the norm $\|\cdot\|_{\mathrm{FBL}[E]}$ will provide the desired representation of FBL$[E]$. Indeed, the process of extension of linear and bounded operators $T : E \to X$, X being a Banach lattice, to lattice homomorphism $\hat{T} : \mathrm{FBL}[E] \to X$ works as follows. First, consider functions $f \in \mathrm{FVL}[E]$, which can be obtained from some evaluation functions $(\delta_{x_i})_{i=1}^n \subset \delta_E(E)$ by means of a finite number of linear and lattice operations, i.e., they can be represented as *lattice-linear expressions* $F(\delta_{x_1}, \dots, \delta_{x_n})$ depending on the $(\delta_{x_i})_{i=1}^n$. Then, defining $\hat{T}f = F(Tx_1, \dots, Tx_n)$ and using lattice-linear function calculus (cf. [67, Section 1.d]) it can be checked that this expression does not depend on the choice of F, so the operator $\hat{T} : \mathrm{FVL}[E] \to X$ is a well-defined lattice homomorphism. Now, the last step in the proof of [21, Theorem 2.5] is to show that $\hat{T}$ is bounded when FVL$[E]$ is endowed with the norm $\|\cdot\|_{\mathrm{FBL}[E]}$, so that it can be uniquely extended to its closure, FBL$[E]$.

The free Banach lattice over a Banach space E can be visualized as a particular case of a more abstract family of free objects generated by Banach spaces in some restrained classes of Banach lattices, the *free $\mathcal{D}$-convex Banach lattices over a Banach space E*, introduced in [57]. A *convexity condition* is a triple $\mathcal{D} = (\mathcal{G}, M, \vartheta)$ consisting of the following elements. $\mathcal{G}$ is a nonempty subset of $\bigcup_{m=1}^{\infty} \mathcal{H}_m^{\geq 0}$, where $\mathcal{H}_m$ denotes the set of all continuous positively homogeneous functions from $\mathbb{R}^m$ to $\mathbb{R}$ and $\mathcal{H}_m^{\geq 0}$ is the subset of the monotone functions of $\mathcal{H}_m$, i.e., the functions $h \in \mathcal{H}_m$ such that $h(t_1, \ldots, t_m) \leq h(s_1, \ldots, s_m)$ whenever $0 \leq t_i \leq s_i$ for every $i = 1, \ldots, m$. On the other hand, $M : \mathcal{G} \to [1, \infty)$ and $\vartheta : \mathcal{G} \to \{0, 1\}$ are just arbitrary functions. Given a convexity condition $\mathcal{D} = (\mathcal{G}, M, \vartheta)$, we say that a seminorm ν over a vector lattice X is *$\mathcal{D}$-convex* if for every $m \in \mathbb{N}$ and $g \in \mathcal{G} \cap \mathcal{H}_m^{\geq 0}$, the inequality

$$\nu(g(x_1, \ldots, x_m)) \leq M(g) g(\nu(x_1), \ldots, \nu(x_m))$$

holds for every $x_1, \ldots, x_m \in X_+$ when $\vartheta(g) = 1$, and for every pairwise disjoint elements $x_1, \ldots, x_m \in X_+$ when $\vartheta(g) = 0$, where the expression $g(x_1, \ldots, x_m)$ must be understood in terms of the lattice-linear function calculus. A Banach lattice X is said to be *$\mathcal{D}$-convex* if its norm is $\mathcal{D}$-convex.

Some examples of convexity conditions are p-convexity and upper p-estimates: Given $1 \leq p \leq \infty$, we say that a Banach lattice X is *p-convex* (respectively, satisfies an *upper p-estimate*) if there exists a constant $M \geq 1$ such that for any finite collection of elements (respectively, pairwise disjoint elements) $x_1, \ldots, x_n \in X$ the inequality

$$\left\| \left(\sum_{i=1}^{n} |x_i|^p \right)^{\frac{1}{p}} \right\| \leq M \left(\sum_{i=1}^{n} \|x_i\|^p \right)^{\frac{1}{p}}$$

holds, making the obvious modifications in the case $p = \infty$. We call p-convexity (respectively, upper p-estimate) constant of X to the least constant M satisfying the above inequality.

We can equip the sublattice FVL$[E] \subset H[E]$ previously defined with the following norm:

$$\|f\|_{\mathrm{FBL}^{\mathcal{D}}[E]} := \sup \left\{ \nu(f) : \nu(\delta_x) \leq \|x\| \ \forall x \in E \right\}$$

for every $f \in \mathrm{FVL}[E]$, where the supremum is taken over all possible $\mathcal{D}$-convex seminorms ν over FVL$[E]$. Then, the completion of FVL$[E]$ under the norm $\|\cdot\|_{\mathrm{FBL}^{\mathcal{D}}[E]}$, denoted $\mathrm{FBL}^{\mathcal{D}}[E]$, is the free $\mathcal{D}$-convex Banach lattice over E [57, Theorem 3.3], i.e., it is a $\mathcal{D}$-convex Banach lattice that satisfies the following universal property: for every $\mathcal{D}$-convex Banach lattice X and every linear and bounded operator $T : E \to X$ there exists a unique lattice homomorphism $\hat{T}$ such that $\hat{T} \circ \delta_E = T$. Moreover, $\|\hat{T}\| = \|T\|$. This procedure provides a family of free objects which includes the free Banach lattice over a Banach space E as a particular case. It suffices to observe that, as a consequence of the triangular inequality, being a Banach lattice is equivalent to being 1-convex with constant 1, which is already a convexity condition. Another relevant case is the *free p-convex Banach lattice over E*, $\mathrm{FBL}^{(p)}[E]$. The existence of this object can be showed in an abstract way by applying the previous procedure with p-convexity as our convexity condition, but there also exists an explicit construction that can be obtained by generalizing the norm in (4.1) to define

$$\|f\|_{\mathrm{FBL}^{(p)}[E]} := \sup\left\{\left(\sum_{i=1}^{m}|f(x_i^*)|^p\right)^{\frac{1}{p}} : m \in \mathbb{N}, (x_i^*)_{i=1}^m \subset E^*, \sup_{x\in B_E}\left(\sum_{i=1}^{m}|x_i^*(x)|^p\right)^{\frac{1}{p}} \leq 1\right\}$$

when $p < \infty$, and

$$\|f\|_{\mathrm{FBL}^{(\infty)}[E]} := \sup_{x^*\in B_{E^*}} |f(x^*)|$$

when $p = \infty$ (see [57, Section 6] and [76, Section 2]).

It is worth mentioning that, even though these previous constructions have been developed in the setting of real Banach spaces and Banach lattices, they can be adapted to the complex setting [54]. For instance, given a complex Banach space E we can obtain the *free complex Banach lattice over E*, $\mathrm{FBL}_{\mathbb{C}}[E]$, through the complexification of an appropriate renorming of a real free Banach lattice. To do so, we consider E as a real Banach space, denoting it by $E_{\mathbb{R}}$, and renorm $\mathrm{FBL}[E_{\mathbb{R}}]$ with

$$\|f\|_{\mathrm{FBL}_{\mathbb{C}}[E]} := \sup\left\{\sum_{i=1}^{m}|f(\mathfrak{Re}\, z_i^*)| : m \in \mathbb{N}, (z_i^*)_{i=1}^m \subset E^*, \sup_{z\in B_E}\sum_{i=1}^{m}|z_i^*(z)| \leq 1\right\},$$

which is equivalent to the original norm $\|\cdot\|_{\mathrm{FBL}[E_{\mathbb{R}}]}$. Then, the complexified Banach lattice $\mathrm{FBL}_{\mathbb{C}}[E] := \mathrm{FBL}[E_{\mathbb{R}}] \oplus i\mathrm{FBL}[E_{\mathbb{R}}]$, endowed with the modulus defined pointwise by

$$|f|(x^*) = \sqrt{f_1(x^*)^2 + f_2(x^*)^2} = |f(x^*)|, x^* \in (E_{\mathbb{R}})^*$$

for every $f = f_1 + if_2 \in \mathrm{FBL}_{\mathbb{C}}[E]$ and the norm

$$\|f\|_{\mathrm{FBL}_{\mathbb{C}}[E]} := \sup\left\{\sum_{i=1}^{m}|f(\mathfrak{Re}\, z_i^*)| : m \in \mathbb{N}, (z_i^*)_{i=1}^m \subset E^*, \sup_{z\in B_E}\sum_{i=1}^{m}|z_i^*(z)| \leq 1\right\},$$

together with the $\mathbb{C}$-linear isometric embedding $\delta_E : E \to \mathrm{FBL}_{\mathbb{C}}[E]$ given by

$$\delta_E(z) = \delta_{E_{\mathbb{R}}}(z) - i\delta_{E_{\mathbb{R}}}(iz),$$

plays the role of the free complex Banach lattice over the complex Banach space E [54, Theorem 3.3].

Finally, note that free Banach lattices can be constructed not only over Banach spaces, but over some other categories too. Consider for example the *free Banach lattice generated by a set A* proposed by B. de Patger and A. W. Wickstead in [79]. It is a Banach lattice $\mathrm{FBL}(A)$ with a bounded map $\iota : A \to \mathrm{FBL}(A)$ such that for every Banach lattice X and every bounded map $T : A \to X$ there exists a unique lattice homomorphism $\hat{T} : \mathrm{FBL}(A) \to X$ with norm $\|\hat{T}\| = \sup\{\|Ta\| : a \in A\}$ that makes the following diagram commute:

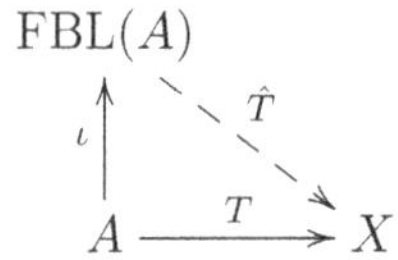

The free Banach lattice over a set happens to be a particular case of the free Banach lattice over a Banach space, since $\mathrm{FBL}(A)$ is lattice isometric to $\mathrm{FBL}[\ell_1(A)]$ [21, Corollary 2.9]. Note that $\ell_1(A)$ can be seen as the free Banach space over the set A, because it satisfies the universal property of uniquely extending every bounded map between A and any Banach space X to a bounded linear operator from $\ell_1(A)$ to X by just defining the operator coordinatewise. Thus, the free Banach lattice over a certain set can be interpreted as the free Banach lattice generated by the free Banach space over this set, providing an example of the interaction of free objects in different categories.

Free Banach lattices generated by a lattice can also be considered. Recall that a *lattice* $\mathbb{L}$ is a partial ordered set that admits two operations of supremum ($\vee$) and infimum ($\wedge$), and a *lattice homomorphism* in this setting is a map between lattices that preserve these two operations. In [22], A. Avilés and J. D. Rodríguez-Abellán showed the existence of the free Banach lattice generated by a lattice $\mathbb{L}$, which is a Banach lattice $\mathrm{FBL}\langle\mathbb{L}\rangle$ together with a bounded lattice homomorphism $\phi : \mathbb{L} \to \mathrm{FBL}\langle\mathbb{L}\rangle$ such that for every Banach lattice X and every bounded lattice homomorphism $T : \mathbb{L} \to X$ there exists a unique Banach lattice homomorphism $\hat{T} : \mathrm{FBL}\langle\mathbb{L}\rangle \to X$ such that $\hat{T} \circ \phi = T$ and $\|\hat{T}\| = \sup\{\|Tx\| : x \in \mathbb{L}\}$. They also provided an explicit construction of $\mathrm{FBL}\langle\mathbb{L}\rangle$, inspired by those of $\mathrm{FBL}(A)$ and $\mathrm{FBL}[E]$. Further developments about this notion can be found in [17,18,23]

4.2 Properties of FBL[*E*]

Let us now focus on some of the properties of $\mathrm{FBL}[E]$. We start by considering the lattice homomorphisms from $\mathrm{FBL}[E]$ to $\mathbb{R}$, which is the same as studying the atoms of $\mathrm{FBL}[E]^*$. These elements are precisely the extensions as lattice homomorphisms of the functionals $x^* \in E^*$, denoted by $\widehat{x^*}$ [21, Corollary 2.7]:

Lemma 4.1. *A functional $\phi \in \mathrm{FBL}[E]^*$ is a lattice homomorphism if and only if there exists some $x^* \in E^*$ such that $\phi = \widehat{x^*}$, i.e., $\phi(f) = f(x^*)$ for every $f \in \mathrm{FBL}[E]$.*

Note that, in particular, $\mathrm{FBL}[E]^*$ is a Banach lattice with many atoms, in contrast with other dual Banach lattices such as $L_p[0,1]$, that has no atoms at all.

Another interesting fact about $\mathrm{FBL}[E]$ is that its pairwise disjoint subsets can be at most countable, no matter which Banach space E we consider [20]. This property is known as the *countable chain condition* (*ccc* for short). In fact, $\mathrm{FBL}[E]$ satisfies a stronger property, the *σ-bounded chain condition* (*σ-bcc*). Both properties can be stated formally as follows.

Definition 4.2. A Banach lattice X

(1) satisfies the *countable chain condition* (*ccc*) if for every uncountable family $\mathcal{F} \subset X_+$ there are distinct $f, g \in \mathcal{F}$ such that $f \wedge g \neq 0$;
(2) satisfies the *σ-bounded chain condition* (*σ-bcc*) if X_+ admits a countable decomposition $X_+ = \bigcup_{n \geq 2} \mathcal{F}_n$ such that, for every n, in every subset $\mathcal{G} \subset \mathcal{F}_n$ of size n there exists two distinct elements $f, g \in \mathcal{G}$ such that $f \wedge g \neq 0$.

In [20], A. Avilés, G. Plebanek and J. D. Rodríguez-Abellán showed that $\mathcal{C}_{ph}(B_{E^*}, \omega^*)$, the space of positively homogeneous and continuous functions over B_{E^*} endowed with the weak* topology, satisfies the σ-bounded chain condition. Since every function in $\mathrm{FBL}[E]$ is the uniform limit in B_{E^*} of functions in $\mathrm{FVL}[E]$, which are piecewise affine and thus weak* continuous on B_{E^*}, it follows that $\mathrm{FBL}[E]$ is a sublattice of $\mathcal{C}_{ph}(B_{E^*}, \omega^*)$. In connection with this, it is shown in [76, Proposition 2.2] that $\mathcal{C}_{ph}(B_{E^*}, \omega^*)$ coincides lattice isometrically with $\mathrm{FBL}^{(\infty)}[E]$.

The case when the underlying Banach space is actually a Banach lattice is of particular interest. First note that in general, for a Banach lattice X, $\mathrm{FBL}[X]$ need not coincide with X. Take for instance the case of a finite dimensional Banach lattice X, where $\mathrm{FBL}[X]$ is lattice isomorphic to $\mathcal{C}(S_{X^*})$ (here, S_{X^*} denotes the unit sphere). This is a common situation in free objects, as the process of embedding X "freely" inside $\mathrm{FBL}[X]$ cannot carry the lattice information of X. Nevertheless, some of it is preserved in a certain sense. For example, when X is a Banach lattice there exists a surjective lattice homomorphism $\beta_X : \mathrm{FBL}[X] \to X$, which can be obtained by applying the universal property of $\mathrm{FBL}[X]$ to the identity operator $id_X : X \to X$. This fact implies that X is complemented (as a Banach space) inside $\mathrm{FBL}[X]$ by means of the projection $\delta_X \circ \beta_X$, whose kernel is an ideal in $\mathrm{FBL}[X]$.

Inspired by the work on Lipschitz free spaces of G. Godefroy and N. J. Kalton [48] (see Sect. 2), the authors of [19] studied when this situation can be improved, so that X can be embedded inside $\mathrm{FBL}[X]$ through a lattice homomorphism $\alpha_X : X \to \mathrm{FBL}[X]$ in such a way that $\beta_X \circ \alpha_X = id_X$. A Banach lattice satisfying the above condition is said to have the *lattice-lifting property*. Some examples of Banach lattices with the lattice-lifting property are projective Banach lattices (see Definition 4.5 below), Banach lattices with a lattice structure induced by a 1-unconditional basis and $\mathrm{FBL}[E]$ for any Banach space E.

A particular aspect in the study of the properties of free Banach lattices is to determine which Banach lattice properties of $\mathrm{FBL}[E]$ encode certain Banach space properties of the generating space E. In [76, Section 9] the authors provide a list of characterizations, some of which are summarized in the following theorem.

Theorem 4.3. *Let E be a Banach space. Then:*

(1) E has finite dimension if and only if $\mathrm{FBL}[E]$ *has a strong unit;*
(2) E is separable if and only if $\mathrm{FBL}[E]$ *has a quasi-interior point;*
(3) E contains a complemented subspace isomorphic to ℓ_1 if and only if $\mathrm{FBL}[E]$ *contains a sublattice isomorphic to ℓ_1;*
(4) The identity operator $id_{E^*} : E^* \to E^*$ is $(p', 1)$-summing if and only if $\mathrm{FBL}[E]$ *satisfies an upper p-estimate, where $1 \leq p \leq \infty$ and $\frac{1}{p} + \frac{1}{p'} = 1$.*

This dictionary between Banach space properties of E and Banach lattice properties of $\mathrm{FBL}[E]$ can also be expanded with some entries related to linear and bounded operators between Banach spaces and the lattice homomorphisms between free Banach lattices generated by them. Given Banach spaces E and F and $T : E \to F$ a linear and bounded operator, we can consider the lattice homomorphism $\overline{T} = \widehat{(\delta_F \circ T)} : \mathrm{FBL}[E] \to \mathrm{FBL}[F]$, which makes the following diagram commutative:

$$\begin{array}{ccc} \mathrm{FBL}[E] & \overset{\overline{T}}{\dashrightarrow} & \mathrm{FBL}[F] \\ {\scriptstyle \delta_E}\uparrow & & \uparrow{\scriptstyle \delta_F} \\ E & \xrightarrow{\;T\;} & F \end{array}$$

Then, T and $\overline{T}$ are related in the following way [76, Section 3.1]:

Theorem 4.4. *Let E and F be Banach spaces and $T : E \to F$ and $\overline{T} : \mathrm{FBL}[E] \to \mathrm{FBL}[F]$ as above. Then:*

(1) $\overline{T}$ is injective if and only if T is injective;
(2) $\overline{T}$ has dense range if and only if T has dense range;
(3) $\overline{T}$ is onto if and only if T is onto;
(4) $\overline{T}$ is an embedding if and only if T is an embedding and there exists some $C > 0$ such that for every $n \in \mathbb{N}$, $\varepsilon > 0$ and $S : E \to \ell_1^n$ there exists an extension $\widetilde{S} : E \to \ell_1^n$ with $\|\widetilde{S}\| \leq C(1 + \varepsilon)\|S\|$.

Note that the previous result implies that if E and F are isomorphic, then $\mathrm{FBL}[E]$ and $\mathrm{FBL}[F]$ are lattice isomorphic. Nevertheless, the reverse question is still open: namely, suppose that $\mathrm{FBL}[E]$ and $\mathrm{FBL}[F]$ are lattice isomorphic, are E and F necessarily linearly isomorphic? Note that in the complex case, this question has a negative answer (see [54]).

4.3 Applications to Banach Lattice Theory

Since they were first proposed in 2015 by B. de Patger and A. W. Wickstead [79], free Banach lattices have been proved to be a very useful tool in Banach lattice theory. Take for instance the case of projectivity in Banach lattices, a concept that was also introduced in [79].

Definition 4.5. A Banach lattice X is called *projective* if there exists some $\lambda \geq 1$ such that whenever Y is a Banach lattice, $J \subset Y$ is a closed ideal and $Q : Y \to Y/J$ is the quotient map, then for every lattice homomorphism $T : X \to Y/J$ there exists a lattice homomorphism $\tilde{T} : X \to Y$ satisfying $Q \circ \tilde{T} = T$ and $\|\tilde{T}\| \leq \lambda \|T\|$.

It turns out that free Banach lattices characterize projective Banach lattices in the following sense [79, Theorem 10.3]:

Theorem 4.6. *A Banach lattice X is projective if and only if for every $\varepsilon > 0$ there is a set A and a sublattice H of* $\mathrm{FBL}(A)$ *such that there exists a lattice isomorphism $\psi : X \to H$ with $\|\psi\|$, $\|\psi^{-1}\| \leq 1+\varepsilon$ and H is lattice complemented in* $\mathrm{FBL}(A)$ *through a projection of norm at most $1+\varepsilon$.*

Further research concerning the relation between projectivity and free Banach lattices can be found in [15].

Other applications of free Banach lattices to Banach lattice theory can be found in [21], where they provide a counterexample to a question raised by J. Diestel on weakly compactly generated Banach lattices. Before formulating the question, let us introduce some key concepts related to the problem. We say that a Banach space X is *weakly compactly generated* (WCG for short) if there exists a weakly compact set $K \subset X$ such that $X = \overline{\mathrm{span}}(K)$. Analogously, a Banach lattice X is said to be *lattice weakly compactly generated* (LWCG) if there is a weakly compact set $L \subset X$ such that the sublattice generated by L is dense in X, i.e., $X = \overline{\mathrm{lat}}(L)$. Clearly, every WCG Banach lattice is LWCG. The question raised by J. Diestel was the converse: is every LWCG Banach lattice WCG? The answer is negative in general, since $\mathrm{FBL}[\ell_2(\Gamma)]$ with Γ uncountable is LWCG (take the image of the unit ball of $\ell_2(\Gamma)$ through the canonical embedding $\delta_{\ell_2(\Gamma)}$ as a weakly compact set that generates the whole $\mathrm{FBL}[\ell_2(\Gamma)]$ as a lattice) but not WCG [21].

An additional open question that has been answered using free Banach lattices comes from [84]. Recall that a sequence of non-zero vectors (x_k) of a Banach lattice is *bibasic* if there exists a constant $M \geq 1$ such that for every $m \in \mathbb{N}$ and scalars $a_1, \ldots, a_m$ the following inequality, called *bibasic inequality*, is satisfied:

$$\left\| \sum_{n=1}^{m} \left| \bigvee_{k=1}^{n} a_k x_k \right| \right\| \leq M \left\| \sum_{k=1}^{m} a_k x_k \right\|.$$

It was asked in [84, Remark 4.4] whether there exists a subspace of a Banach lattice with no bibasic sequences. In [76, Theorem 7.5], T. Oikhberg, M. A. Taylor, P. Tradacete and V. G. Troitsky showed that the canonical copy $\delta_{c_0}(c_0)$ in $\mathrm{FBL}[c_0]$ is a subspace without a bibasic sequence, providing a positive answer to the previous question.

Free Banach lattices have also been used to prove the existence of *push-outs* in the category of Banach lattices. In category theory, given three objects X_0, X_1 and X_2 and two morphisms $\alpha_i : X_0 \to X_i$, $i = 1, 2$, a *push-out* (also known as *amalgamated sum* or *fibered coproduct*) is an object PO together with two morphism $\beta_i : X_i \to PO$, $i = 1, 2$, such that the diagram

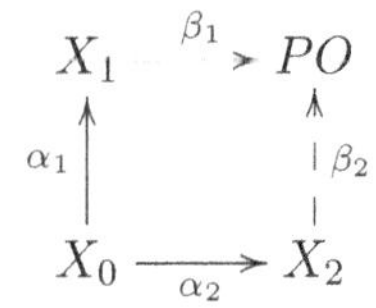

commutes and the following universal property is satisfied: if there exists another object B and two morphisms $\beta_i' : X_i \to B$, $i = 1, 2$, such that $\beta_1' \circ \alpha_1 = \beta_2' \circ \alpha_2$, then there is a unique $\gamma : PO \to B$ such that $\beta_i' = \gamma \circ \beta_i$ for $i = 1, 2$, as the following diagram illustrates.

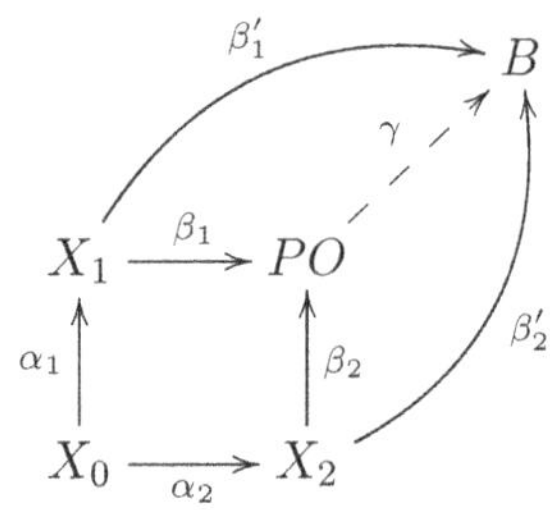

In the category of Banach lattices with lattice homomorphisms, we can also define an *isometric push-out* as a push-out such that $\|\beta_i\| \leq 1$, $i = 1, 2$, and $\|\gamma\| \leq \max\left\{\|\beta_1'\|, \|\beta_2'\|\right\}$ in the universal property. In [24, Theorem 4.3] A. Avilés a P. Tradacete proved the existence of isometric push-outs using the free Banach lattice generated by a Banach space as an auxiliary tool. Moreover, they proved the following embedding property as a corollary (see also [85]).

Theorem 4.7. *Given $T_1 : X_0 \to X_1$ and $T_2 : X_0 \to X_2$ isometric lattice embeddings between Banach lattices, then there exists a Banach lattice X_3 and isometric lattice embeddings $S_i : X_i \to X_3$, $i = 1, 2$, such that $S_1 \circ T_1 = S_2 \circ T_2$.*

Free Banach lattices have also been used in [37] to provide the first example of a lattice homomorphism not attaining its norm.

References

1. Albiac, F., Ansorena, J.L., Cúth, M., Doucha, M.: Embeddability of ℓ_p and bases in Lipschitz free p-spaces for $0<p \leq 1$. J. Funct. Anal. **278**(4), 108354 (2020). 33 pp
2. Albiac, F., Ansorena, J.L., Cúth, M., Doucha, M.: Lipschitz free p-spaces for $0<p<1$. Israel J. Math. **240**(1), 65–98 (2020)
3. Albiac, F., Ansorena, J.L., Cúth, M., Doucha, M.: Lipschitz free spaces isomorphic to their infinite sums and geometric applications. Trans. Amer. Math. Soc. **374**(10), 7281–7312 (2021)

4. Albiac, F., Kalton, N.J.: Lipschitz structure of quasi-Banach spaces. Israel J. Math. **170**, 317–335 (2009)
5. Aliaga, R.: Extreme points in Lipschitz-free spaces over compact metric spaces. Mediterr. J. Math. **19**(1) (2022). Paper No. 32, 12 pp
6. Aliaga, R.J., Gartland, C., Petitjean, C., Procházka, A.: Purely 1-unrectifiable spaces and locally flat Lipschitz functions. Trans. Amer. Math. Soc. **375**, 3529–3567 (2022)
7. Aliaga, R.J., Guirao, A.J.: On the preserved extremal structure of Lipschitz-free spaces. Stud. Math. **245**, 1–14 (2019)
8. Aliaga, R.J., Pernecká, E.: Supports and extreme points in Lipschitz-free spaces. Rev. Mat. Iberoam. **36**(7), 2073–2089 (2020)
9. Aliaga, R.J., Pernecká, E., Petitjean, C., Procházka, A.: Supports in Lipschitz-free spaces and applications to extremal structure. J. Math. Anal. Appl. **489**(1), 124128 (2020)
10. Aliaga, R.J., Petitjean, C., Procházka, A.: Compact reduction in Lipschitz free spaces. Studia Math. **260**(3), 341–359 (2021)
11. Aliaga, R.J., Petitjean, C., Procházka, A.: Embeddings of Lipschitz-free spaces into ℓ_1. J. Funct. Anal. **280**(6), 108916 (2021)
12. Ando, T.: On the predual of H^∞. Comment Math. Special Issue 1, 33–40 (1978). Special issue dedicated to Władysław Orlicz on the occasion of his seventy-fifth birthday
13. Arazy, J.: An application of infinite dimensional holomorphy to the geometry of banach spaces. In: Lindenstrauss, J., Milman, V.D. (eds.) Geometrical Aspects of Functional Analysis. LNM, vol. 1267, pp. 122–150. Springer, Heidelberg (1987). https://doi.org/10.1007/BFb0078141
14. Aron, R., Dimant, V., García-Lirola, L.C., Maestre, M.: Linearization of holomorphic Lipschitz functions. Math. Nachr. https://doi.org/10.1002/mana.202300527
15. Avilés, A., Martínez-Cervantes, G., Rodríguez-Abellán, J.D.: On projective Banach lattices of the form $C(K)$ and FBL[E]. J. Math. Anal. Appl. **489**(1), 124129 (2020). 11 pp
16. Avilés, A., Martínez-Cervantes, G., Rueda Zoca, A.: Local complementation under free constructions. Quaestiones Math. https://arxiv.org/abs/2107.11339
17. Avilés, A., Martínez-Cervantes, G., Rodríguez-Abellán, J.D., Rueda Zoca, A.: Free Banach lattices generated by a lattice and projectivity. Proc. Amer. Math. Soc. **150**, 2071–2082 (2022)
18. Avilés, A., Martínez-Cervantes, G., Rodríguez-Abellán, J.D., Rueda Zoca, A.: Lattice embeddings in free Banach lattices over lattices. Math. Inequal. Appl. **25**(2), 495–509 (2022)
19. Avilés, A., Martínez-Cervantes, G., Rodríguez, J., Tradacete, P.: A Godefroy-Kalton principle for free Banach lattices. Israel J. Math. **247**, 433–458 (2022)
20. Avilés, A., Plebanek, G., Rodríguez-Abellán, J.D.: Chain conditions in free Banach lattices. J. Math. Anal. Appl. **465**(2), 1223–1229 (2018)
21. Avilés, A., Rodríguez, J., Tradacete, P.: The free Banach lattice generated by a Banach space. J. Funct. Anal. **274**(10), 2955–2977 (2018)
22. Avilés, A., Rodríguez-Abellán, J.D.: The free Banach lattice generated by a lattice. Positivity **23**(3), 581–597 (2019)
23. Avilés, A., Rodríguez-Abellán, J.D.: Projectivity of the free Banach lattice generated by a lattice. Arch. Math. (Basel) **113**(5), 515–524 (2019)
24. Avilés, A., Tradacete, P.: Amalgamation and injectivity in Banach lattices. Int. Math. Res. Not. IMRN **2**, 956–997 (2023)

25. Baker, K.A.: Free vector lattices. Canad. J. Math. **20**, 58–66 (1968)
26. Benyamini, Y., Lindenstrauss, J.: Geometric Nonlinear Functional Analysis, vol. 1. American Mathematical Society Colloquium Publications, p. 48. American Mathematical Society (2000)
27. Bleier, R.D.: Free vector lattices. Trans. Amer. Math. Soc. **176**, 73–87 (1973)
28. Boyd, C.: Some topological properties of Preduals of spaces of holomorphic functions. Proc. Royal Irish Acad. Sect. A Math. Phys. Sci. **94A**(2), 167–178 (1994)
29. Çalişkan, E.: Bounded holomorphic mappings and the compact approximation property in Banach spaces. Port. Math. **61**(1), 25–33 (2004)
30. Çalişkan, E.: Approximation of holomorphic mappings on infinite dimensional spaces. Rev. Mat. Complut. **17**(2), 411–434 (2004)
31. Çalişkan, E.: The bounded approximation property for the predual of the space of bounded holomorphic mappings. Studia Math. **177**, 225–233 (2006)
32. Carando, D., Zalduendo, I.: Linearization of functions. Math. Ann. **328**, 683–700 (2004)
33. Cheeger, J., Kleiner, B., Naor, A.: Compression bounds for Lipschitz maps from the Heisenberg group to L_1. Acta Math. **207**(2), 291–373 (2011)
34. Cúth, M., Doucha, M.: Lipschitz-free spaces over ultrametric spaces. Mediterr. J. Math. **13**(4), 1893–1906 (2016)
35. Cúth, M., Doucha, M., Wojtaszczyk, P.: On the structure of Lipschitz-free spaces. Proc. Am. Math. Soc. **144**(9), 3833–3846 (2016)
36. Daniilidis, A., Sepulcre, J.M., Venegas, F.M.: Asymmetric free spaces and canonical asymmetrizations. Studia Math. **261**(1), 55–102 (2021)
37. Dantas, S., Martínez-Cervantes, G., Rodríguez Abellán, J., Rueda Zoca, A.: Norm-attaining lattice homomorphisms. Rev. Mat. Iberoam. **38**(3), 981–1002 (2022)
38. Dineen, S.: Complex Analysis in Locally Convex Spaces. North.-Holland Math. Studies, vol. 57. North-Holland, Amsterdam (1981)
39. Dutrieux, Y., Ferenczi, V.: The Lipschitz free Banach spaces of $C(K)$-spaces. Proc. Amer. Math. Soc. **134**(4), 1039–1044 (2006)
40. Fabian, M., Habala, P., Hájek, P., Montesinos, V., Zizler, V.: Banach Space Theory. The Basis for Linear and Nonlinear Analysis. CMS Books in Mathematics/Ouvrages de Mathématiques de la SMC. Springer, New York (2011). https://doi.org/10.1007/978-1-4419-7515-7
41. Galindo, P., García, D., Maestre, M.: Holomorphic mappings of bounded type. J. Math. Anal. Appl. **166**(1), 236–246 (1992)
42. Galindo, P., Miralles, A.: Asplund Sets and Metrizability for the Polynomial Topology. J. Convex Anal. **18**(2), 433–446 (2011)
43. García-Lirola, L., Petitjean, C., Procházka, A.: Injectivity of Lipschitz operators. Bull. Malays. Math. Sci. Soc. **46** (2023). Paper no. 68
44. García-Lirola, L., Petitjean, C., Procházka, A., Rueda Zoca, A.: Extremal structure and duality of Lipschitz free spaces. Mediterr. J. Math. **15**(2) (2018). Paper no. 69, 23 pp
45. García-Lirola, L., Procházka, A., Rueda Zoca, A.: A characterisation of the Daugavet property in spaces of Lipschitz functions. J. Math. Anal. Appl. **464**(1), 473–492 (2018)
46. Godard, A.: Tree metrics and their Lipschitz-free spaces. Proc. Amer. Math. Soc. **138**, 4311–4320 (2010)
47. Godefroy, G.: A survey on Lipschitz-free Banach spaces. Comment. Math. **55**(2), 89–118 (2015)
48. Godefroy, G., Kalton, N.J.: Lipschitz-free Banach spaces. Studia Math. **159**, 121–141 (2003)

49. Godefroy, G., Ozawa, N.: Free Banach spaces and the approximation properties. Proc. Am. Math. Soc. **142**(5), 1681–1687 (2014)
50. Gupta, M., Baweja, D.: Weighted spaces of holomorphic functions on Banach spaces and the approximation property. Extracta Math. **31**(2), 123–144 (2016)
51. Gupta, M., Baweja, D.: The compact approximation property for weighted spaces of holomorphic mappings. In: Ruzhansky, M., Cho, Y.J., Agarwal, P., Area, I. (eds.) Advances in Real and Complex Analysis with Applications. TM, pp. 33–47. Springer, Singapore (2017). https://doi.org/10.1007/978-981-10-4337-6_2
52. Hájek, P., Lancien, G., Pernecká, E.: Approximation and Schur properties for Lipschitz free spaces over compact metric spaces. Bull. Belg. Math. Soc. Simon Stevin **23**(1), 63–72 (2016)
53. Hájek, P., Pernecká, E.: On Schauder bases in Lipschitz-free spaces. J. Math. Anal. Appl. **416**(2), 629–646 (2014)
54. de Hevia, D., Tradacete, P.: Free complex Banach lattices. J. Funct. Anal. **284**(10), 109888 (2023). 26 pp
55. Isbell, J.R.: Some remarks concerning categories and subspaces. Can. J. Math. **9**, 563–577 (1957)
56. Ivakhno, Y., Kadets, V., Werner, D.: The Daugavet property for spaces of Lipschitz functions. Math. Scand. **101**(2), 261–267 (2007). Corrigendum ibid. 104 (2009), No. 2, 319
57. Jardón-Sánchez, H., Laustsen, N.J., Taylor, M.A., Tradacete, P., Troitsky, V.G.: Free Banach lattices under convexity conditions. RACSAM Rev. R. Acad. Cienc. Exactas Fís. Nat. Ser. A Mat. **116**(1) (2022). Paper no. 15
58. Jiménez-Vargas, A., Ruiz-Casternado, D., Sepulcre, J.M.: On holomorphic mappings with compact type range. Bull. Malays. Math. Sci. Soc. **46**, 22 (2023). Paper no. 20
59. Jiménez-Vargas, A., Saadi, K., Sepulcre, J.M.: Cohen strongly p-summing holomorphic mappings on Banach spaces. Banach J. Math. Anal. **17**, 44 (2023). https://arxiv.org/abs/2209.03038
60. Jung, M.: Daugavet property of Banach algebras of holomorphic functions and norm-attaining holomorphic functions. Adv. Math. **421**, 109005 (2023)
61. Kalton, N.J.: Spaces of Lipschitz and Hölder functions and their applications. Collect. Math. **55**(2), 171–217 (2004)
62. Kaufmann, P.L.: Products of Lipschitz-free spaces and applications. Studia Math. **226**(3), 213–227 (2015)
63. Kaup, W., Upmeier, H.: Banach spaces with biholomorphically equivalent unit balls are isomorphic. Proc. Amer. Math. Soc. **58**, 129–133 (1978)
64. Kochanek, T., Pernecká, E.: Lipschitz-free spaces over compact subsets of superreflexive spaces are weakly sequentially complete. Bull. Lond. Math. Soc. **50**(4), 680–696 (2018)
65. Lancien, G., Pernecká, E.: Approximation properties and Schauder decompositions in Lipschitz-free spaces. J. Funct. Anal. **264**(10), 2323–2334 (2013)
66. Lindenstrauss, J., Tzafriri, L.: Classical Banach Spaces I. Springer, Heidelberg (1977)
67. Lindenstrauss, J., Tzafriri, L.: Classical Banach Spaces II. Springer, Heidelberg (1979)
68. MacLane, S.: Duality for groups. Bull. Am. Math. Soc. **56**, 485–516 (1950)
69. Mazet, P.: Analytic Sets in Locally Convex Spaces, North-Holland Math. Stud., vol. 89. North-Holland, Amsterdam (1984)
70. Meyer-Nieberg, P.: Banach Lattices. Springer, Heidelberg (1991)

71. Mujica, J.: Linearization of bounded holomorphic mappings on Banach spaces. Trans. Amer. Math. Soc. **324**(2), 867–887 (1991)
72. Mujica, J.: Linearization of holomorphic mappings on infinite dimensional spaces. Rev. Un. Mat. Argentina **37**, 127–134 (1991)
73. Mujica, J.: Linearization of holomorphic mappings of bounded type. In: Progress in Functional Analysis. North-Holl. Math. Stud., vol. 170, pp. 149–162. North-Holland, Amsterdam (1992)
74. Mujica, J., Nachbin, L.: Linearization of holomorphic mappings on locally convex spaces. J. Math. Pures Appl. **71**(6), 543–560 (1992)
75. Naor, A.: An introduction to the Ribe program. Jpn. J. Math. **7**(2), 167–233 (2012)
76. Oikhberg, T., Taylor, M.A., Tradacete, P., Troitsky, V.G.: Free Banach lattices. J. Eur. Math. Soc. (JEMS). https://arxiv.org/abs/2210.00614
77. Ostrovska, S., Ostrovskii, M.I.: Generalized transportation cost spaces. Mediterr. J. Math. **16**(6) (2019). Paper no. 157
78. Ostrovska, S., Ostrovskii, M.I.: On relations between transportation cost spaces and ℓ_1. J. Math. Anal. Appl. **491**(2), 124338 (2020). 12 pp
79. de Pagter, B., Wickstead, A.W.: Free and projective Banach lattices. Proc. Roy. Soc. Edinburgh Sect. A **145**(1), 105–143 (2015)
80. Pernecká, E., Smith, R.J.: The metric approximation property and Lipschitz-free spaces over subsets of $\mathbb{R}^N$. J. Approx. Theory **199**, 29–44 (2015)
81. Petitjean, C.: Lipschitz-free spaces and Schur properties. J. Math. Anal. Appl. **453**(2), 894–907 (2017)
82. Ravichandran, M., Yavuz, O.: Preduals of H^∞ of finitely connected domains. Proc. Amer. Math. Soc. **142**(5), 1641–1648 (2014)
83. Semadeni, Z.: Free and direct objects. Bull. Am. Math. Soc. **69**(1), 63–66 (1963)
84. Taylor, M.A., Troitsky, V.G.: Bibasic sequences in Banach lattices. J. Funct. Anal. **278**(10), 108448 (2020). 33 pp
85. Tursi, M.A.: A separable universal homogeneous Banach lattice. Int. Math. Res. Not. IMRN **7**, 5438–5472 (2023)
86. Weaver, N.: Lipschitz Algebras. World Scientific (1999)
87. Weaver, N.: On the unique predual problem for Lipschitz spaces. Math. Proc. Camb. Phil. Soc. **165**(3), 467–473 (2018)
88. Wojtaszczyk, P.: On projections in spaces of bounded analytic functions with applications. Studia Math. **65**, 147–173 (1979)

Orthogonal Polynomials and Operator Theory. Open Problems and Applications

Edmundo J. Huertas[1] and Francisco Marcellán[2(✉)]

[1] Departamento de Física y Matemáticas, Facultad de Ciencias, Universidad de Alcalá, Ctra. Madrid-Barcelona, Km. 33,600, 28805 Alcalá de Henares, Madrid, Spain
edmundo.huertas@uah.es

[2] Departamento de Matemáticas, Universidad Carlos III de Madrid, Avenida Universidad, 30, 28911 Leganés, Madrid, Spain
pacomarc@ing.uc3m.es

Abstract. In this contribution we review first the connections between orthogonal polynomials on the real line, Jacobi matrices, their LU and UL factorizations, spectral measures, spectral transformations and Stieltjes functions as well as their application in the study of some integrable systems of the Toda hierarchy. In a second step, we analyze symmetric operators associated with a Sobolev type inner product. The connection between the Cholesky factorization of five diagonal matrices associated with such operators and the UL factorization of the square of a shifted Jacobi matrix associated with a Christoffel perturbation of the measure is pointed out. Finally, we deal with orthogonal polynomials associated with probability measures supported on the unit circle, Hessenberg and CMV matrices, spectral measures and their spectral transformations, Carathéodory functions as well as their application in the analysis of Schur flows both from an algebraic and analytic way.

AMS Subject Classification (2000): Primary 33C47 · Secondary 42C05

1 Introduction

The aim of this work is to present the close relationship between orthogonal polynomials (OP in the sequel) and operator theory, both on the real line and on the unit circle, as well as to survey the spectral transformations and various representations which are recent research objects. We will briefly review the state-of-the-art at OP and spectral theory, including some illustrative applications derived of the research on these issues.

In the first part, we will discuss the basic elements of OP theory with respect to nontrivial probability measures supported on the real line (OPRL, in short): the multiplication operator, Jacobi matrices and their spectral resolution, the Stieltjes function, as well as Darboux's linear and rational spectral transformations. This introduction will serve to motivate the two models of integrable

H. Herrero et al. (Eds.): RSME 2022, RSME 13, pp. 125–152, 2024.
https://doi.org/10.1007/978-3-031-62025-6_7

systems that we present. On the one hand, we will solve the so-called Uvarov flow, connected with the linear spectral transformations of the same name, and, on the other hand, we will analyze a well-known system, the Toda flow, of which we will give the analytical solution in the finite and infinite cases. On the other hand, we will focus our attention on a symmetric operator with respect to a Sobolev-type inner product whose representation in terms of the corresponding family of orthonormal polynomials is a five diagonal matrix. Once the Cholesky factorization of such a matrix is implemented, the commutation of factors yields the square of a shifted Jacobi matrix corresponding to a Christoffel perturbation of the measure involved in the Sobolev inner product.

Finally, we study OP sequences defined by a nontrivial probability measure supported on the unit circle (OPUC, in short). They illustrate inner products such that the multiplication operator is a unitary operator. We will analyze integrable systems that are related to two representations of the multiplication operator, namely the so-called GGT representation (Geronimus, Gragg and Teplyaev) and the CMV representation (Cantero, Moral and Velázquez). There is much less known here about the integrable system models discussed in the case of the real line. The only model systematically studied is the so-called Schur flow, which constitutes the final part of the manuscript.

The relation between the solutions of the full Kostant-Toda lattice and the discrete Korteweg-de Vries equation is analyzed in [4]. A method for constructing solutions of these systems is given. As a consequence of the matricial interpretation of this method, the Darboux's linear transformation is extended for general Hessenberg banded matrices. On the other hand, in [5] the relation between the concept of Darboux transform and the full Kostant Toda lattice is analyzed. The discrete Korteweg de Vries equation is used to obtain new solutions of the full Kostant Toda lattice. In addition, an iterative method to obtain the generalized Darboux factorization for a Hessenberg banded matrix is provided. In such a way new solutions are deduced.

In a more general framework we must point out that the Gauss-Borel or LU factorization of Gram matrices associated with bilinear forms which are no necessarily symmetric or Hermitian constitutes a key element in the theory of biorthogonal polynomials. The construction of sequences of biorthogonal polynomials and its second kind functions, the spectral matrices modeling the multiplication by the independent variable x, the Christoffel-Darboux kernel and its projection properties, are discussed in [21].

2 Orthogonal Polynomials on the Real Line

2.1 The Multiplication Operator

Let $\mathbb{P}$ be the linear space of polynomials with real coefficients and $\mathbb{P}_n \subset \mathbb{P}$ the linear subspace of polynomials of degree at most n. We begin by describing the representation of the multiplication operator, which in the case of OP with real coefficients, is the multiplication by the variable x for each polynomial. We will denote this operator by $M_x : \mathbb{P} \to \mathbb{P}$, that is,

$$M_x(p(x)) = xp(x) \quad \text{for every } p \in \mathbb{P}.$$

Let us consider in $\mathbb{P}$ an ***inner product*** $\langle \cdot, \cdot \rangle : \mathbb{P} \times \mathbb{P} \to [0, \infty)$, such that M_x is a symmetric operator with respect to it, i.e.,

$$\langle M_x p(x), q(x) \rangle = \langle p(x), M_x q(x) \rangle .$$

The most common case is to impose that this inner product is positive definite, and then there is at least one positive spectral measure μ, supported on the real line, such that the inner product can be represented as

$$\langle p, q \rangle = \int_{\mathbb{R}} pq d\mu, \quad p, q \in \mathbb{P}.$$

Let μ be a nontrivial probability measure, supported on an infinite subset E on the real line. Under these conditions, there is a sequence of polynomials $\{p_n(x)\}_{n\geq 0}$, with $deg p_n = n, n \geq 0$, such that

$$\int_E p_n(x) p_m(x) d\mu(x) = \delta_{n,m}.$$

The system $\{p_n(x)\}_{n\geq 0}$ is said to be an ***orthonormal polynomial sequence*** with respect to the measure μ. It is very well known that this sequence satisfies the following ***three term recurrence relation (3TRR)***

$$xp_n(x) = a_{n+1} p_{n+1}(x) + b_n p_n(x) + a_n p_{n-1}(x), \quad n \geq 0, \tag{1}$$

where $a_n > 0$, with initial conditions $p_{-1}(x) = 0$, and $p_0(x) = 1$.

The above recurrence relation can be also expressed in a matrix form

$$x\mathbf{p}(x) = \bar{\mathbf{J}}\mathbf{p}(x),$$

where $\mathbf{p}(x) = [p_0(x), p_1(x), p_2(x), \ldots]^T$ and $\bar{\mathbf{J}}$ is called the ***Jacobi tridiagonal matrix***

$$\bar{\mathbf{J}} = \begin{bmatrix} b_0 & a_1 & 0 & \cdots \\ a_1 & b_1 & a_2 & \cdots \\ 0 & a_2 & b_2 & \cdots \\ \vdots & \vdots & \vdots & \ddots \end{bmatrix}.$$

Normalizing the polynomials of the above sequence in such a way that their *leading coefficient* (that is, the coefficient of the term with the highest power) equals to 1, the polynomial sequence is denoted $\{P_n(x)\}_{n\geq 0}$, and is called a ***monic orthogonal polynomial sequence***. In this case the 3TRR (1) becomes

$$xP_n(x) = P_{n+1}(x) + \beta_n P_n(x) + \gamma_n P_{n-1}(x), \quad n \geq 0,$$

Notice that $\beta_n = b_n$ and $\gamma_n = a_n^2$. Thus, the matrix representation of the multiplication operator in terms of the monic OPRL is

$$\mathbf{J} = \begin{bmatrix} \beta_0 & 1 & 0 & \cdots \\ \gamma_1 & \beta_1 & 1 & \cdots \\ 0 & \gamma_2 & \beta_2 & \cdots \\ \vdots & \vdots & \vdots & \ddots \end{bmatrix},$$

which is called now a ***monic Jacobi matrix***. We must point out the fact that the eigenvalues of the leading principal submatrix of size $n \times n$ of $\mathbf{J}$ are the zeros of the polynomial $P_n(x)$ (see [12]).

Likewise, there exist another important objects which are the functions

$$S_n(x) = \int_E \frac{p_n(t)d\mu(t)}{x-t}, \tag{2}$$

known as the ***second kind functions associated with*** μ. They constitute a second independent solution of the difference equation

$$xy_n = a_{n+1}y_{n+1} + b_ny_n + a_ny_{n-1}, \quad n \geq 0,$$

satisfied by the OPRL.

When $n = 0$ in (2), we get the ***Stieltjes function*** associated with the measure μ, that is the Cauchy transform of the measure

$$S(x) = S_0(x) = \int_E \frac{d\mu(t)}{x-t}.$$

The Stieltjes function S_0 plays an important role in Approximation Theory, because the approximation of this function around the point of infinity by rational functions (Padé approximation), gives rise to denominators that are precisely the OPRL associated with the measure μ, and the polynomials in the numerator of the approximant are the so-called OPRL of the second kind.

The Stieltjes function has the following Laurent expansion at infinity

$$S(x) = \frac{\mu_0}{x} + \sum_{k=1}^{\infty} \frac{\mu_k}{x^{k+1}},$$

where μ_k are the ***moments*** of the measure μ, i.e.,

$$\mu_k = \int_E x^k d\mu(x).$$

Taking this into account, we assume μ is a nontrivial probability measure, and then we have $\mu_0 = 1$. Notice that the Stieltjes function of the measure μ is the z-transform of the sequence of the moments, i.e., it is the "mirror" of such a sequence. On the other hand, there is a connection between Stieltjes functions and continued fractions associated with the coefficients of the 3TRR (see [12]).

2.2 Spectral Transformation

We will now analyze some transformations corresponding to modifications of a measure μ.

2.2.1 Canonical Linear Spectral Transformations

Given a measure μ supported on $\mathbb{R}$, three canonical models of ***linear spectral transformations on the real line*** are considered in the literature. Namely,

1. Christoffel transformation ($\mathcal{R}_C$), multiplication by a positive polynomial of degree 1
$$d\tilde{\mu} = (x - \beta)d\mu, \quad \beta \notin supp(\mu). \tag{3}$$
2. Geronimus transformation ($\mathcal{R}_G$), division of the measure by a positive polynomial of degree 1, plus a Dirac delta supported on a point of the real line that does not belong to the support of μ
$$d\hat{\mu} = \frac{d\mu}{x - \beta} + M\delta(x - \beta), \quad \beta \notin supp(\mu), M \in \mathbb{R}_+. \tag{4}$$
3. Uvarov transformation ($\mathcal{R}_U$), addition of a Dirac delta supported on a point of the real line
$$d\breve{\mu} = d\mu + M\delta(x - \beta), \quad M \in \mathbb{R}_+. \tag{5}$$

We can establish relationships between them. First, the composition of the Geronimus transformation, and the Christoffel transformation based in the same point yields the identity transformation $\mathcal{I}$

$$\mathcal{R}_C \circ \mathcal{R}_G = \mathcal{I}.$$

On the other hand, the composition of the Christoffel transformation and the Geronimus transformation leads to an Uvarov transformation

$$\mathcal{R}_G \circ \mathcal{R}_C = \mathcal{R}_U.$$

2.2.2 Rational Spectral Transformations

Another model of spectral transformations on the real line are the so called ***rational spectral transformations***. Given a family of orthonormal polynomials $\{p_n(x)\}_{n\geq 0}$, we can shift the indices in the recurrence relation, and then we obtain the so called ***associated monic polynomials of k-th kind*** ($\mathcal{R}^{(k)}$), which are the sequence of polynomials $\{p_n^{(k)}(x)\}_{n\geq 0}$ satisfying the following 3TRR

$$xp_n^{(k)}(x) = a_{n+k+1}p_{n+1}^{(k)}(x) + b_{n+k}p_n^{(k)}(x) + a_{n+k}p_{n-1}^{(k)}(x), \quad n \geqslant 0,$$

with $p_{-1}^{(k)}(x) = 0$ and $p_0^{(k)}(x) = 1$.

In a similar way, by shifting down k rows and columns in the monic Jacobi matrix associated with the corresponding measure, they arise the so-called ***anti-associated monic polynomials of order k*** ($\mathcal{R}^{(-k)}$)

$$xp_n^{(-k)}(x) = \tilde{a}_{n+1}p_{n+1}^{(-k)}(x) + \tilde{b}_n p_n^{(-k)}(x) + \tilde{a}_n p_{n-1}^{(-k)}(x), \quad n \geq 0,$$

with

$$p_{-1}^{(k)}(x) = 0, \quad p_0^{(k)}(x) = 1,$$
$$\{\tilde{b}_n\}_{n\geq 0} = \{\hat{b}_i\}_{i=0}^{k-1} \bigcup \{b_n\}_{n\geq 0},$$
$$\{\tilde{a}_n\}_{n\geq 1} = \{\hat{a}_i\}_{i=1}^{k} \bigcup \{a_n\}_{n\geq 1},$$

where $\{\hat{b}_i\}_{i=0}^{k-1}$ and $\{\hat{a}_i\}_{i=1}^{k}$ are free parameters.

The composition of the above transformations generates the identity transformation

$$\mathcal{R}^{(k)} \circ \mathcal{R}^{(-k)} = \mathcal{I}.$$

2.2.3 Spectral Transformations and Stieltjes Functions

In the spectral transformations studied above, the descriptors ***linear*** and ***rational*** come basically from the behavior of the Stieltjes functions, which represent their corresponding measures. In the first case, the relationship between the corresponding Stieltjes functions is linear, in the sense that

$$\widetilde{S}(x) = \frac{A(x)S(x) + B(x)}{D(x)},$$

where $A(x)$, $B(x)$ and $D(x)$ are polynomials, while in the second case, their relationship is of rational type

$$\widetilde{S}(x) = \frac{A(x)S(x) + B(x)}{C(x)S(x) + D(x)}, \quad AD - BC \neq 0.$$

In 1997, A. Zhedanov [34] established that any kind of rational spectral transformation can be considered generated by Christoffel ($\mathcal{R}_C$), Geronimus ($\mathcal{R}_G$) and the rational transformations ($\mathcal{R}^{(k)}$) and ($\mathcal{R}^{(-k)}$).

2.2.4 Matrix Darboux Transformations

What interpretation do we give from the matrix point of view (i.e., from the point of view of the operator), to the above spectral transformations? The answer is directly related to the matrix representation of the multiplication operator. Having $\mathbf{J}$, the monic Jacobi matrix associated with a nontrivial probability measure μ supported on the real line, and assuming that all the main leading principal submatrices of the shifted Jacobi matrix $\mathbf{J}$ are nonsingular, there exists a unique factorization of the form

$$\mathbf{J} = \mathbf{LU},$$

with

$$\mathbf{L} = \begin{bmatrix} 1 & 0 & 0 & \cdots \\ \ell_1 & 1 & 0 & \cdots \\ 0 & \ell_2 & 1 & \cdots \\ \vdots & \vdots & \vdots & \ddots \end{bmatrix}, \quad \mathbf{U} = \begin{bmatrix} u_1 & 1 & 0 & \cdots \\ 0 & u_2 & 1 & \cdots \\ 0 & 0 & u_3 & \cdots \\ \vdots & \vdots & \vdots & \ddots \end{bmatrix},$$

where, due to the special structure of **J**, the lower triangular matrix **L**, normalized with 1 as diagonal entries, is a lower bidiagonal matrix, while the **U** is an upper bidiagonal matrix. Now we reverse the order of the factorization, constructing the **UL** matrix, and this is in turn a new Jacobi matrix, which we call $\widetilde{\mathbf{J}}$. The transformation that passes from **J** to $\widetilde{\mathbf{J}}$ is called a ***Darboux transformation without parameter***, and $\widetilde{\mathbf{J}} = \mathbf{UL}$ is called the ***Darboux transformation of* J *without parameter*** (see [7]).

On the basis of the above, we can reinterpret the meaning of the canonical Christoffel transformation (3). At a point $x = \beta$ outside the support of the measure, the evaluation of the corresponding OPRL is different from zero. In fact, a necessary and sufficient condition for the existence of the factorization **LU** of $\mathbf{J} - \beta\mathbf{I}$ is that

$$P_n(\beta) \neq 0, \quad \forall n \geq 1.$$

Next, we shift the x variable, and we construct the shifted matrix $\mathbf{J} - \beta\mathbf{I}$ which is still a Jacobi matrix. We first consider its **LU** factorization, next we reverse the order by constructing **UL**, and therefore we obtain the matrix $\widetilde{\mathbf{J}} - \beta\mathbf{I}$, which is the monic Jacobi matrix associated with the Christoffel transformation. Consequently, the canonical Christoffel transformation in terms of spectral theory, is nothing else than the multiplication by the polynomial $(x - \beta)$, and thus $\widetilde{\mathbf{J}}$ is the monic Jacobi matrix corresponding to $d\tilde{\mu} = (x - \beta)d\mu$.

Proposition 1 (Canonical Christoffel transformation). *If $P_n(\beta) \neq 0$, $n \geq 1$, then*

$$\mathbf{J} - \beta\mathbf{I} = \mathbf{LU},$$
$$\widetilde{\mathbf{J}} - \beta\mathbf{I} = \mathbf{UL}.$$

In the same way, there is a non-unique factorization which allows the matrix **J** to be represented as a product (in this precise order) of $\tilde{\mathbf{U}}$ by $\tilde{\mathbf{L}}$. Here $\tilde{\mathbf{L}}$ is lower bidiagonal with ones as diagonal entries, and $\tilde{\mathbf{U}}$ is upper bidiagonal and has a parameter of this non-unique operation as $(1, 1)$ entry. Defining now the reverse order of that matrix product $\tilde{\mathbf{L}}\tilde{\mathbf{U}}$, we have as well another tridiagonal matrix, in this case known as the ***Darboux transformation of* J *with parameter*** (see [7]).

Thus, if now we first consider the $\tilde{\mathbf{U}}\tilde{\mathbf{L}}$ factorization of the $\mathbf{J} - \beta\mathbf{I}$ matrix, and next we reverse the order of factors by constructing $\tilde{\mathbf{L}}\tilde{\mathbf{U}}$, we obtain a monic Jacobi matrix as well, which is the monic Jacobi matrix associated with the canonical Geronimus transformation. In this case, from the point of view of the corresponding spectral measure, we have a Geronimus transformation (4).

Proposition 2 (Canonical Geronimus transformation). *If $\beta \notin supp(\mu)$, then*

$$\mathbf{J} - \beta\mathbf{I} = \tilde{\mathbf{U}}\tilde{\mathbf{L}},$$
$$\widehat{\mathbf{J}} - \beta\mathbf{I} = \tilde{\mathbf{L}}\tilde{\mathbf{U}},$$

where $\widehat{\mathbf{J}}$ is the monic Jacobi matrix corresponding to $d\hat{\mu} = \dfrac{d\mu}{x - \beta} + M\delta(x - \beta)$.

For the Uvarov transformation the idea is very similar, considering that this is the composition of a canonical Christoffel transformation and a canonical Geronimus transformation. The mass appears as a kind of side effect, as a consequence of the fact that the UL factorization we are dealing with is not unique.

Proposition 3 (Canonical Uvarov transformation). *If*

$$\begin{aligned} \mathbf{J} - \beta\mathbf{I} &= \mathbf{L}_1\mathbf{U}_1, \\ \mathbf{J}_1 - \beta\mathbf{I} &= \mathbf{U}_1\mathbf{L}_1 = \mathbf{U}_2\mathbf{L}_2, \\ \check{\mathbf{J}} - \beta\mathbf{I} &= \mathbf{L}_2\mathbf{U}_2, \end{aligned}$$

then $\check{\mathbf{J}}$ *is the monic Jacobi matrix for the measure* $d\check{\mu} = d\mu + M\delta(x-\beta)$*, where* $M = \dfrac{\mu_0(b_0 - \beta - s)}{s}$ *and* s *is the free parameter corresponding to the UL factorization of* $\mathbf{J}_1$.

For more information about linear spectral transformations, Jacobi matrices, orthogonal polynomials and Stieltjes functions, see [15,31] and [34].

2.3 Sobolev-Type Transformations of a Measure

In this section we will focus our attention on the following Sobolev-type inner product

$$\langle f, g\rangle_S = \int_E f(x)g(x)d\mu + Mf(c)g(c) + Nf'(\beta)g'(\beta), \quad f,\, g \in \mathbb{P}, \tag{6}$$

where μ is a positive Borel measure supported on $E = [a,b] \subseteq \mathbb{R}$, $\beta \notin E$, and M, N are nonnegative real numbers with $N > 0$. In general, E can be a bounded or unbounded interval on the real line. Let $\{s_n^{M,N}(x)\}_{n\geq 0}$ denote the orthonormal polynomial sequence with respect to (6). These polynomials are known in the literature as *Sobolev-type* or *discrete Sobolev* orthogonal polynomials. It is worth to point out that many properties of the standard orthogonal polynomials are lost when an inner product as (6) is considered. In the above cases, the three-term recurrence relation that every sequence of orthogonal polynomials (with respect to a measure supported on an infinite subset of the real line) satisfies, does not hold anymore. This is a direct consequence of the fact that the multiplication operator by x is not symmetric with respect to any of the above mentioned situations. Indeed, it is well known that polynomials in $\{s_n^{M,N}(x)\}_{n\geq 0}$ satisfy higher order recurrence relations, and we would show here that there is an intimate relation between these higher order recurrence relations and the canonical spectral transformations on the real line.

We first note that the Christoffel canonical transformation of a measure μ, given in (3), can be iterated in the following way

$$d\tilde{\mu}^{[k]} = (x-\beta)^k d\mu, \quad k \geq 0, \tag{7}$$

and the corresponding system is said to be the k-iterated Christoffel orthogonal polynomial sequence with respect to the above measure. This standard family satisfies a three term recurrence relation for any $k \geq 0$, and for the specific case when $k = 2$, the orthonormal polynomial sequence $\{p_n^{[2]}(x)\}_{n\geq 0}$ satisfies the three term recurrence relation

$$xp_n^{[2]}(x) = a_{n+1}^{[2]}p_{n+1}^{[2]}(x) + b_n^{[2]}p_n^{[2]}(x) + a_n^{[2]}p_{n-1}^{[2]}(x), \quad n \geq 0,$$

with $p_{-1}^{[2]}(x) = 0$, $p_0^{[2]}(x) = 1/\sqrt{a_0^{[2]}}$, where $a_n^{[2]}$ and $b_n^{[2]}$ are known sequences (see [19]).

In this section we analyze the higher order recurrence relation that a sequence of Sobolev-type orthonormal polynomials satisfies when one considers the orthonormal family $\{s_n^{M,N}(x)\}_{n\geq 0}$ with respect to the inner product (6), and the orthonormal family associated with the measures $d\mu$ and $(x-\beta)^2 d\mu$, respectively. A matrix analysis of the five diagonal symmetric matrix associated with such a higher order recurrence relation is presented, taking into account the QR factorization of the shifted symmetric Jacobi matrix related to the polynomials orthonormal with respect to the measure $d\mu$. The shifted Jacobi matrix associated with $(x-\beta)^2 d\mu$ is RQ (see [9] and the references therein).

As we have seen in the previous section, the connection formulas are the main tool to study the analytic properties of new families of OPS, in terms of other families of OPS with well-known analytic properties. Indeed, the problem of finding such expressions is called ***the connection problem***, and it is of great importance in this context.

In the sequel, the following notation will be useful. For every $k \in \mathbb{N}_0$, let us define $\mathbf{J}_{[k]}$ as the semi-infinite symmetric Jacobi matrix associated with the measure $(x-\beta)^k d\mu$, such that

$$x\,\mathbf{p}^{[k]} = \mathbf{J}_{[k]}\,\mathbf{p}^{[k]},$$

where $\mathbf{p}^{[k]}$ denotes the semi-infinite column vector with *orthonormal* polynomial entries $\mathbf{p}^{[k]} = [p_0^{[k]}(x), p_1^{[k]}(x), p_2^{[k]}(x), \ldots]^\top$, being $\{p^{[k]}(x)\}_{n\geq 0}$ the orthonormal polynomial sequence with respect to the modified measure defined in (7). Notice that $\mathbf{p}^{[0]} = \mathbf{p} = [p_0(x), p_1(x), p_2(x), \ldots]^\top$, being $\{p(x)\}_{n\geq 0}$ the orthonormal polynomial sequence with respect to the standard measure μ, and $\mathbf{J}_{[0]} = \mathbf{J}$ is the corresponding Jacobi matrix.

Let $\{p_n^{[2]}(x)\}_{n\geq 0}$ be the sequence of polynomials orthonormal with respect to the modified measure (7) with $k = 2$. Then the following connection formula holds

$$s_n^{M,N}(x) = \gamma_{n,n}p_n^{[2]}(x) + \gamma_{n-1,n}p_{n-1}^{[2]}(x) + \gamma_{n-2,n}p_{n-2}^{[2]}(x), \tag{8}$$

where $\gamma_{n,n}$, $\gamma_{n-1,n}$, and $\gamma_{n-1,n}$ are known sequences (see [19]).

We observe that the multiplication operator by $(x-\beta)^2$ is a symmetric operator with respect to the discrete Sobolev inner product (6). In other words, for any $p(x), q(x) \in \mathbb{P}$, it fulfills

$$\langle (x-\beta)^2 p(x), q(x)\rangle_S = \langle p(x), (x-\beta)^2 q(x)\rangle_S.$$

Therefore the sequence $\{s_n^{M,N}(x)\}_{n\geq 0}$ satisfies a five-term recurrence relation for $n \geq 0$ as follows

$$(x-\beta)^2 s_n^{M,N}(x) =$$
$$a_{n+2}s_{n+2}^{M,N}(x) + b_{n+1}s_{n+1}^{M,N}(x) + c_n s_n^{M,N}(x) + b_n s_{n-1}^{M,N}(x) + a_n s_{n-2}^{M,N}(x), n \geq 0,$$

with $s_{-2}^{M,N}(x) = s_{-1}^{M,N}(x) = 0$.

2.3.1 A Matrix Approach

It is possible to obtain a very interesting relation between the five diagonal matrix $\mathbf{H}$, related to the multiplication operator by $(x-\beta)^2$, the corresponding Sobolev-type orthonormal polynomials, and the Jacobi matrix $\mathbf{J}_{[2]}$ associated with the 2–iterated orthonormal polynomials $\{p_n^{[2]}(x)\}_{n\geq 0}$.

First, we deal with the matrix representation of the above five-term recurrence relation satisfied by the sequence of Sobolev-type orthonormal polynomials $\{s_n^{M,N}(x)\}_{n\geq 0}$

$$(x-\beta)^2 \mathbf{s}^{M,N} = \mathbf{H}\,\mathbf{s}^{M,N},$$

where $\mathbf{H}$ is the five diagonal semi-infinite symmetric matrix

$$\mathbf{H} = \begin{bmatrix} c_0 & b_1 & a_2 & 0 & \cdots \\ b_1 & c_1 & b_2 & a_3 & \cdots \\ a_2 & b_2 & c_2 & b_3 & \cdots \\ 0 & a_3 & b_3 & c_3 & \cdots \\ \vdots & \vdots & \vdots & \vdots & \ddots \end{bmatrix}, \tag{9}$$

and $\mathbf{s}^{M,N} = [s_0^{M,N}(x), s_1^{M,N}(x), s_2^{M,N}(x), \ldots]^{\intercal}$. Next, the matrix version of the connection formula (8) reads

$$\mathbf{s}^{M,N} = \mathbf{T}\,\mathbf{p}^{[2]},$$

where $\mathbf{T}$ is the banded lower triangular, semi-infinite and nonsingular matrix with positive diagonal entries

$$\mathbf{T} = \begin{bmatrix} \gamma_{0,0} & 0 & 0 & 0 & \cdots \\ \gamma_{0,1} & \gamma_{1,1} & 0 & 0 & \cdots \\ \gamma_{0,2} & \gamma_{1,2} & \gamma_{2,2} & 0 & \cdots \\ 0 & \gamma_{1,3} & \gamma_{2,3} & \gamma_{3,3} & \cdots \\ \vdots & \vdots & \vdots & \vdots & \ddots \end{bmatrix} \tag{10}$$

and $\mathbf{p}^{[2]} = [p_0^{[2]}(x), p_1^{[2]}(x), p_2^{[2]}(x), \ldots]^{\intercal}$.

Under these conditions, it can be proved (see [19]) that the semi-infinite five diagonal matrix $\mathbf{H}$, can be obtained from the matrix $(\mathbf{J}_{[2]} - c\mathbf{I})^2$ as follows

$$\mathbf{H} = \mathbf{T}(\mathbf{J}_{[2]} - \beta\mathbf{I})^2\mathbf{T}^{-1}.$$

Moreover, it can be shown that the following factorization

$$\mathbf{J} - \beta\mathbf{I} = \mathbf{QR}$$

holds. Here $\mathbf{R}$ is a upper triangular matrix, and $\mathbf{Q}$ is an orthogonal matrix, i.e. $\mathbf{QQ}^\top = \mathbf{Q}^\top\mathbf{Q} = \mathbf{I}$, and swapping the order of these two last semi-infinite matrices, one finds

$$\mathbf{RQ} = \mathbf{J}_{[2]} - \beta\mathbf{I}$$

where $\mathbf{J}_{[2]}$ is the symmetric Jacobi matrix such that

$$x\,\mathbf{p}^{[2]} = \mathbf{J}^{[2]}\mathbf{p}^{[2]}.$$

We emphasize that the above provides an alternative proof of Theorem 3.3 in [9].

Since $\mathbf{J}_{[2]}$ is a symmetric matrix, and $\mathbf{QQ}^\top = \mathbf{I}$, we easily observe

$$(\mathbf{J}_{[2]} - \beta\mathbf{I})^2 = (\mathbf{J}_{[2]} - \beta\mathbf{I})(\mathbf{J}_{[2]} - \beta\mathbf{I})^\top = \mathbf{RQQ}^\top\mathbf{R}^\top = \mathbf{RR}^\top.$$

Thus, we also find that the square of the positive definite symmetric matrix $\mathbf{J}_{[2]} - \beta\mathbf{I}$ has the following factorization

$$(\mathbf{J}_{[2]} - \beta\mathbf{I})^2 = \mathbf{RR}^\top,$$

where $\mathbf{R}$ is a upper triangular matrix. Furthermore

$$(\mathbf{J} - \beta\mathbf{I})^2 = \mathbf{R}^\top\mathbf{R}.$$

Indeed, it can be proved that the five diagonal semi-infinite symmetric matrix $\mathbf{H}$ defined in (9), has the following Cholesky factorization

$$\mathbf{H} = \mathbf{TT}^\top,$$

where $\mathbf{T}$ is the banded, lower triangular, semi-infinite matrix defined in (10). Besides, for any positive Borel measure $d\mu$ supported on an infinite subset $E \subseteq \mathbb{R}$, if $\mathbf{J}$ is the corresponding semi-infinite symmetric Jacobi matrix, and assuming that β does not belong to the interior of the convex hull of E, then for the 2–iterated perturbed measure $(x - \beta)^2 d\mu$ such that $\mathbf{J}_{[2]}$ is the corresponding semi-infinite symmetric Jacobi matrix, we get

$$(\mathbf{J}_{[2]} - \beta\mathbf{I})^2 = \mathbf{RR}^\top = \mathbf{T}^\top\mathbf{T}.$$

For a careful analysis of these questions, see [19].

2.4 Integrable Systems

Although in physics and mathematics we can find few different notions of the concept, we can say that an ***integrable system*** is essentially a dynamical system, whose solutions analytically depend on the initial conditions (see [8]). The first example we will analyze here is the so called ***Uvarov flow***, and we will see that, indeed, it is related with the Uvarov transformation (5). The second is a well-known integrable system called the ***Toda flow***.

2.4.1 Uvarov Flow

We begin with a sequence of OPRL with respect to an even measure (supported in a symmetric interval). In this case, the OPRL satisfies a 3TRR with $b_n = 0$, $n \geq 0$, (see [12]). As a consequence, the corresponding Jacobi matrix has zeros in its main diagonal. We denote by u_n the coefficients of p_{n+1} and p_{n-1} in (1). We now suppose that the sequence $\{u_n\}_{n\geq 0}$ depends on a time parameter, i.e., $u_n = u_n(t)$, and applying the 3TRR we obtain an OPRL sequence that will depend on both variables x and t, that is, $P_n(x) = P_n(x;t)$. Taking derivatives with respect to the time variable we have

$$\dot{P}_n(x;t) = C_{n-2}^{(n)}(t)P_{n-2}(x;t) + \cdots + C_{n-2k}^{(n)}(t)P_{n-2k}(x;t) + \cdots, \tag{11}$$

where the coefficients $C_k^{(n)}$ define the time dynamics of $u_n(t)$ and $P_n(x;t)$. In (11), $\dot{P}_n$ is expressed as a linear combination of all the preceding polynomials up to degree $n-2$. In other words, these coefficients define the time dynamics of the system through $u_n(t)$ and therefore, its influence on the polynomials.

Next we study two particular integrable cases, determining the value of the previous coefficients, as well as the spectral measure corresponding to the sequence of orthogonal polynomials $\{P_n(x;t)\}_{n\geq 0}$ for each integrable system, see [33].

The Volterra Chain

First, we will assume that all the coefficients in (11) are zero, except the first one, i.e.,

$$\dot{P}_n(x;t) = C_{n-2}^{(n)}(t)P_{n-2}(x;t).$$

Then $C_{n-2}^{(n)}(t) = -u_n(t)u_{n-1}(t)$ and the following nonlinear system comes into play

$$\dot{u}_n(t) = u_n(t)\left(u_{n+1}(t) - u_{n-1}(t)\right).$$

This is known in the literature as the ***Volterra flow*** or ***Volterra chain***. This system does not respond to any of the spectral transformations that we have previously discussed. We find that the system (the spectral measure) evolves with time from the initial condition, through a perturbation of type $\exp(tx^2)$

$$\omega(x;t) = K(t)\exp\left(tx^2\right)\omega(x;0),$$

where $K(t)$ is the normalization constant in order to $\omega(x;t)$ be a probability measure.

In a more general way, if in (11) $C_{n-2k}^{(n)} \neq 0$ for some $k > 1$, one obtains the nonlinear equation

$$\dot{u}_n(t) = F(u_{n-k}(t), \ldots, u_{n+k}(t))$$

and the problem to obtain here the orthogonality measure $d\mu(x,t)$ is still an open problem.

Uvarov Flow and Christoffel-Darboux Formula

Let us now see another example, a priori far from the analyzed above. We know that for this system, the time derivative of the odd polynomials is zero, and the time derivative of the even polynomials can be expressed as follows

$$\begin{aligned}\dot{P}_{2n+1}(x;t) &= 0,\\ \dot{P}_{2n}(x;t) &= r_n x^{-1} P_{2n-1}(x;t).\end{aligned} \tag{12}$$

Next we recall a classical result for OPRL sequences the ***Christoffel-Darboux formula*** (see [12]). The polynomial $x^{-1}P_{2n-1}(x;t)$ can be expressed as a linear combination of all even polynomials, with $k = 0, \ldots, n-1$. We then have an expansion where all the polynomials of the sequence appear

$$\dot{P}_{2n}(x;t) = \sum_{k=0}^{n-1} C_{2k}^{(n)}(t) P_{2k}(x;t)$$

(let us remind that in the Volterra case it only appears one term). Now the coefficients $C_{2k}^{(n)}(t)$ in the above linear combination are fixed in terms of the initial conditions. In this particular case, unlike the Volterra model, a system of differential equations appears. There the derivatives of u_n are given in terms of certain parameters r_n, which are in turn functions of u_n (the parameters of the 3TRR) and we remember they are our unknowns

$$r_n = r_1 \frac{u_3 u_5 \cdots u_{2n-1}}{u_2 u_4 \cdots u_{2n-2}},$$

also verifying that the derivatives of the odd order terms are given as r_{n+1}.

The integrability of the system (12) is a straightforward consequence of the fact that the spectral parameter x does not depend on t. The coefficients are

$$C_{2k}^{(n)}(t) = \frac{r_n h_{2n-2} P_{2k}(0)}{h_{2k} P_{2n-2}(0)},$$

where

$$\begin{aligned}P_{2n}(0) &= (-1)^n u_1 u_3 \cdots u_{2n-1},\\ h_k &= \|P_k\|^2 = u_1 u_2 \cdots u_k.\end{aligned}$$

They are known as the non-local equations for the integrable system

$$\left.\begin{aligned}\dot{u}_{2n} &= r_n\\ \dot{u}_{2n+1} &= -r_{n+1}\end{aligned}\right\}.$$

There are infinite constants of motion of the type

$$I_k = \operatorname{Trace}(L^k),$$

where the difference operator L is defined by

$$\begin{aligned}L\,|0\rangle &= u_1\,|1\rangle\\ L\,|n\rangle &= u_{n+1}\,|n+1\rangle + |n-1\rangle\end{aligned}$$

for some basis $|n\rangle$, with $n = 0, 1, 2, \ldots$

Proposition 4 **([33]).** *Let assume $\omega(x;t_0) \neq 0$. Then the spectral measure for the sequence $\{P_n(x;t)\}_{n\geq 0}$ has the following dependence on time*

$$\omega(x;t) = (1 - J(t))\omega(x;t_0) + J(t)\delta(x).$$

Here $J(t) = 1 - exp(-\int_{t_0}^{t} \frac{ds}{u_1(s)})$.

Notice that $\omega(x;t)$, the spectral measure of the $\{P_n(x;t)\}_{n\geq 0}$, is a Uvarov perturbation at $x = 0$ of the initial weight where the mass evolves with the time.

2.4.2 Finite Toda Flow

The second model we are dealing with is a well known integrable system. In 1975, J. Moser [22] proposed a method for the integration of a finite Toda flow. In 1986, Yu. M. Berezanskii [6] extended the Moser's method to semi-infinite Toda flows. The problem is the following: we have a Schrödinger operator $T(t)$, with a potential function $q(x,t)$, and another operator $B(t)$ defined by

$$\begin{aligned} T(t) &= -\frac{\partial^2}{\partial x^2} + q(x,t), \\ B(t) &= -4\frac{\partial^3}{\partial x^3} + 3\left(q(x,t)\frac{\partial}{\partial x} + \frac{\partial}{\partial x}q(x,t)\right). \end{aligned}$$

Then, considering the Lax pair (B,T) and denoting by $[B,T]$ the commutator of them, we get

$$\frac{dT}{dt} = [B,T]. \tag{13}$$

The functions that define this potential, which are compatible with the above Lax pair, are those that satisfy the well-known ***Korteweg-de Vries*** (KdV) equation,

$$q_t = 6qq_x - q_{xxx}$$

with the appropriate boundary conditions for q (B must be anti-symmetric and T symmetric). Basically what happens is that the potential $q(x,t)$ satisfies a KdV equation, which is a partial derivative equation that includes non-linearity and dispersion effects, while being associated with a representation in terms of Lax pairs of Lax, in terms of commutation.

Finite Toda Flow

From the point of view of the classical mechanics, this system is known as the Toda lattice, which in the finite case is a lattice with a finite number of oscillators in every node of the lattice. From a differential point of view, the ***finite Toda flow*** is a chain of oscillators that, instead of having an elastic interaction between them given by Hooke's Law, their interaction is of exponential type $e^{-(x_k - x_{k-1})}$, where $(x_k - x_{k-1})$ is the distance between the positions of the nodes (the oscillators). We impose the following boundary conditions on the one-dimensional lattice

$$\begin{aligned} \ddot{x}_k &= \exp(-(x_k - x_{k-1})) - \exp(-(x_{k+1} - x_k)),\ k = 1, 2, \ldots, n, \\ x_0 &= -\infty, \quad x_{n+1} = \infty. \end{aligned} \tag{14}$$

By making a change of variable, the second order differential system is transformed into a first order differential system. For this we introduce the functions

$$\begin{aligned} a_k(t) &= \tfrac{1}{2}\exp\tfrac{1}{2}(x_k(t) - x_{k+1}(t)), \; k = 1, 2, \ldots, n-1, \\ b_k(t) &= -\tfrac{1}{2}\dot{x}_k(t), \qquad\qquad\qquad\quad\; k = 1, 2, \ldots, n. \end{aligned} \tag{15}$$

When using them in (14), they give rise to the first order nonlinear system of differential equations with polynomial type representation

$$\begin{aligned} \dot{a}_k(t) &= a_k(t)\,(b_{k+1}(t) - b_k(t)), \; b_k(0) = \beta_k, \; k = 1, \ldots, n, \\ \dot{b}_k(t) &= 2\left(a_k^2(t) - a_{k-1}^2(t)\right), \quad a_k(0) = \alpha_k, \; k = 1, \ldots, n-1, \end{aligned}$$

with $a_0 \equiv a_n \equiv 0$, which, through the first derivatives, provides the time evolution of the oscillators.

To deal with the above system through the OPRL theory, we built the finite Jacobi matrix $T(t)$ of size $n \times n$ associated with the previously defined functions in (15), and the finite anti-symmetric matrix $B(t)$

$$T(t) = \begin{bmatrix} b_1 & a_1 & & \\ a_1 & b_2 & \ddots & \\ & \ddots & \ddots & a_{n-1} \\ & & a_{n-1} & b_n \end{bmatrix}, \; B(t) = \begin{bmatrix} 0 & a_1 & & \\ -a_1 & 0 & \ddots & \\ & \ddots & \ddots & a_{n-1} \\ & & -a_{n-1} & 0 \end{bmatrix}.$$

Thus, the time derivative of the Jacobi matrix is simply the commutator of $T(t)$ and the antisymmetric matrix $B(t)$, as in (13)

$$\dot{T}(t) = [B, T].$$

Let us now analyze the spectral measure corresponding to $T(t)$. The spectrum $\sigma[T(t)]$ of the $n \times n$ matrix $T(t)$ is the same as the spectrum of $T(t)$ at time zero, i.e., the spectrum of $T(0)$. Thus, this is an isospectral flow, or in other words, the evolution of the eigenvalues of this matrix is time-independent, and also the $\lim_{t\to\pm\infty} T(t)$ of this tridiagonal matrix will be a diagonal matrix, whose entries in the main diagonal are the eigenvalues of the Jacobi matrix where we have the initial conditions. Concerning the corresponding orthogonality measure (the spectral measure), developing in continued fractions, as is done for the Stieltjes functions, we can verify and using the OPRL theory, that this representation in continued fraction comes as in *(iii)* of the following proposition (see Theorem 5.16 and 5.17 in [13] as well as Theorem 10.3 in [26]).

Proposition 5. *The following statements hold*

(i) $\sigma[T(t)] = \sigma[T(0)]$.
(ii) $\lim_{t\to\pm\infty} T(t) = diag(\lambda_1, \lambda_2, \ldots, \lambda_n)$, *where* $(\lambda_k)_{k=1}^n$ *are the eigenvalues of* $T(0)$.

(iii) The solutions of the finite Toda lattice satisfy

$$\cfrac{1}{z-b_1-\cfrac{a_1^2}{z-b_2-\cfrac{a_2^2}{\ddots\,\cfrac{a_{n-1}^2}{z-b_n}}}}=\frac{\sum_{k=1}^{n} m_k \frac{\exp(-\lambda_k t)}{z-\lambda_k}}{\sum_{k=1}^{n} m_k \exp(-\lambda_k t)},$$

where $m_k = \left[\sum_{j=0}^{n} p_j(\lambda_k)^2\right]^{-1} = \dfrac{1}{K_n(\lambda_k,\lambda_k)}$.

Observe that this expression is the Stieltjes transform of a discrete measure, which corresponds to the perturbation by $\exp(-xt)$ of the discrete measure supported in the spectrum of the matrix with weights m_k, $k = 1, 2, \ldots, n$, at the points λ_k, $k = 1, 2, \ldots, n$, of its support.

Let us analyze this from an algebraic point of view, by using QR factorization, see [13]. If $X \in \mathbb{C}^{(n,n)}$, we denote by X_- the strict lower triangular component of X and let $\Pi_0(X) = X_- - X_-^T$ be the anti-symmetric projection of X. We can consider a representation of the Lax pair as follows

$$\dot{T}(t) = [B, T],$$

where $B = -\Pi_0(F(t))$ and F is a function $C^{(k)}(\mathbb{R})$ such that $F(T(0)) = F_0$. Taking into account the QR factorization of the exponential matrix $\exp(tF_0) = Q(t)R(t)$, it can be proved that the solution comes from $T(0)$, a unitary matrix which appears in the factorization $Q(t)$, and its transpose $Q^T(t)$. Hence, one has the following result (see Theorem 5.18 in [13]).

Theorem 1. *If $F(T(0)) = F_0$, let $\exp(tF_0) = Q(t)R(t)$ be its corresponding QR factorization. Then $T(t) = Q^T(t)T(0)Q(t)$ is the solution of the equation $\dot{T}(t) = [B, T]$.*

This result is consistent with the fact that, since $Q(t)$ is unitary, the eigenvalues of $T(0)$ and $T(t)$ are time independent.

When you deal with the function $F(x) = \ln x$ the solution to the Toda problem, the so called QR flow, can be given in an iterative process through the QR algorithm. Indeed, if $T(0)$ is positive definite, let $T_0 = T(0) = QR$, $T_1 = RQ$, $T_2 = Q_1R_1, \ldots$ be the sequence of matrices generated by the QR iteration, and denote $T(k) = T_k$. Thus you can recover the solution of the QR flow at a finite integer time.

The role of the theory of OPRL sequences in the analysis of the finite Toda lattice has been emphasized in [20] and [26], among others.

2.4.3 Semi-infinite Toda Flow

Let now study the case when we have an infinite number of oscillators in the lattice. In this case, the representation of the spectral measure is not going to be through a discrete measure, but through a continuous one. We begin with the same idea as above regarding the initial conditions, and we also give the solution, see Theorem 10.4 in [26],

Theorem 2 ([26]). *Let $\{a_n(t)\}_{n\geq 1}$, $\{b_n(t)\}_{n\geq 0}$ be sequences of real functions. The following statements are equivalent:*

(i) $\{a_n(t)\}_{n\geq 1}$, $\{b_n(t)\}_{n\geq 0}$ *satisfy the equations of the Toda flow*

$$\begin{aligned} \dot{a}_n(t) &= a_n(t)\left(b_{n+1}(t) - b_n(t)\right), & b_n(0) &= \beta_n, \\ \dot{b}_n(t) &= 2\left(a_n^2(t) - a_{n-1}^2(t)\right), & a_n(0) &= \alpha_n, \end{aligned}$$

with $\sup\{|\alpha_n|, |\beta_n|\} < +\infty$.

(ii) The Jacobi semi-infinite matrix $T(t)$ associated with the sequences $\{a_n(t)\}_{n\geq 1}$, $\{b_n(t)\}_{n\geq 0}$ satisfies the Lax equation $\dot{T}(t) = [B, T]$, where B is a semi-infinite antisymmetric matrix with entries $B_{j,j+1} = a_j(t), j = 1, 2, \cdots, B_{j+1,j} = -a_j(t), j = 1, 2, \cdots$ and 0, otherwise.

(iii) The orthogonality measure $\mu(x,t)$ such that $\{a_n(t)\}_{n\geq 1}$, $\{b_n(t)\}_{n\geq 0}$ give us the entries of the corresponding semi-infinite Jacobi matrix is

$$d\mu(x,t) = C(t)\exp(-xt)\,d\mu(x,0),$$

where $C(t)$ is a normalization coefficient in order to μ be a probability measure. Indeed, $\{a_n(t)\}_{n\geq 1}$, $\{b_n(t)\}_{n\geq 0}$ are the coefficients of the continued fraction in the variable z associated with the function $\frac{\int_S e^{-\lambda t}(z-\lambda)^{-1}d\mu(\lambda,0)}{\int_S e^{-\lambda t}d\mu(\lambda,0)}$, where S denotes the support of the spectral measure $d\mu(\lambda,0)$ of the semi-infinite Jacobi matrix $T(0)$.

The connection between OPRL, Toda latices and Stieltjes functions has been deeply analyzed in [27]. On the other hand, an application of the discrete time Toda lattice to the progressive Lanczos algorithm has been pointed out in [24].

2.4.4 Generalization of the Toda Flows Through Freud-Type Perturbations

In [3] the dynamics of the evolution of the measures that are an extension of what we saw for the Volterra case are studied. More precisely, a perturbation of a measure at time equal to zero by

$$d\mu(x,t) = C(t)\exp\left(-x^{2n}t\right)d\mu(x,0)$$

and therefore this transformation of the initial conditions, leads to nonlinear systems in terms of the parameters of the 3TRR.

3 Orthogonal Polynomials on the Unit Circle

In the case of unit circle, much less is known about the type of integrable systems than those analyzed in the case of the real line. The model that has been studied in more depth is the one called ***Schur flow***. First, we introduce some basics facts of orthogonal polynomials on the unit circle (OPUC, in short).

3.1 Moment Functional and Orthogonal Polynomials in $\mathbb{T}$

Let $\mathcal{L}$ be a linear functional in the linear space $\Lambda = span\{z^n\}_{n\in\mathbb{Z}}$ of Laurent polynomials with complex coefficients, such that $\mathcal{L}$ is Hermitian

$$c_n = \langle \mathcal{L}, z^n\rangle = \overline{\langle \mathcal{L}, z^{-n}\rangle} = \bar{c}_{-n}, \quad n \in \mathbb{Z}.$$

Then we can introduce a bilinear functional associated with $\mathcal{L}$ in the linear space $\mathbb{P}$ of polynomials with complex coefficients, such that

$$\langle p(z), q(z)\rangle_{\mathcal{L}} = \left\langle \mathcal{L}, p(z)\overline{q}(z^{-1})\right\rangle,$$

where $p, q \in \mathbb{P}$, and $\{c_n\}_{n\geq 0}$ is called the ***moment sequence associated with*** $\mathcal{L}$.

Notice that $\langle zp(z), zq(z)\rangle = \langle p(z), q(z)\rangle$ for $p, q \in \mathbb{P}$.

The ***Gram matrix*** (moment matrix) associated with this bilinear functional, in terms of the monomial basis $\{z^n\}_{n\geq 0}$ of $\mathbb{P}$ is

$$\mathbf{T} = \begin{bmatrix} c_0 & c_{-1} & \cdots & c_{-n} & \cdots \\ c_1 & c_0 & \cdots & c_{-(n-1)} & \cdots \\ \vdots & \vdots & \ddots & \vdots & \\ c_n & c_{n-1} & \cdots & c_0 & \cdots \\ \vdots & \vdots & & \vdots & \ddots \end{bmatrix}.$$

This semi-infinite matrix is known in the literature as hermitian ***Toeplitz matrix***. The functional $\mathcal{L}$ is called quasi-definite if $\mathbf{T}_n$, the leading principal submatrix of size $(n+1)\times(n+1)$, is nonsingular for every $n \geq 0$. In this case, there exists a ***sequence of monic polynomials*** $\{\Phi_n\}_{n\geq 0}$ with $\deg \Phi_n = n$, such that

$$\langle \Phi_n, \Phi_m\rangle_{\mathcal{L}} = \mathbf{k}_n\delta_{n,m}.$$

In the sequel, we will denote this sequence as monic OPUC. It satisfies the following forward (resp. backward) recurrence formulas

$$\Phi_{n+1}(z) = z\Phi_n(z) + \Phi_{n+1}(0)\Phi_n^*(z),$$
$$\Phi_{n+1}(z) = (1 - |\Phi_{n+1}(0)|^2)z\Phi_n(z) + \Phi_{n+1}(0)\Phi_{n+1}^*(z),$$

where $\Phi_n^*(z) = z^n\bar{\Phi}_n(z^{-1})$ is called the ***reverse polynomial.***

The complex numbers $\{\Phi_n(0)\}_{n\geq 1}$ are known in the literature as ***reflection parameters***, ***Schur parameters***, or most commonly ***Verblunsky coefficients***. If $\mathbf{T}_n$ has positive determinant for every $n \geq 0$, then $\mathcal{L}$ is said to be positive definite. In this case, we have the following integral representation

$$\langle \mathcal{L}, p(z) \rangle = \int_{\mathbb{T}} p(z) d\sigma(z),$$

where σ is a nontrivial positive Borel measure, supported on the unit circle $\mathbb{T} = \{z \in \mathbb{C} : |z| = 1\}$. Therefore, there exists a unique orthonormal polynomial sequence $\{\varphi_n\}_{n\geq 0}$ with respect to $\mathcal{L}$ such that

$$\langle \varphi_n, \varphi_m \rangle_{\mathcal{L}} = \delta_{n,m}.$$

Note that, in this case, we have $|\Phi_n(0)| < 1, \;\; \forall n \geq 1$. In the quasi-definite case, where $\det \mathbf{T}_n \neq 0, \;\; n \geq 0$, we have $|\Phi_n(0)| \neq 1, \;\; \forall n \geq 1$. There is an analogous theorem to the Favard's on the real line, Indeed,

Theorem 3 ([28]). *If $\{\alpha_n\}_{n\geq 1}$ is a sequence of complex numbers such that $|\alpha_n| < 1, n = 1, 2, \ldots$, then there exists a unique nontrivial probability measure σ supported on the unit circle, such that α_n are the Verblunsky coefficients $\{\Phi_n(0)\}_{n\geq 1}$ associated with the measure σ. This measure can be decomposed in an absolutely continuous part and in a singular part as follows*

$$d\sigma(\theta) = f(\theta)\frac{d\theta}{2\pi} + d\sigma_s(\theta).$$

The measure σ belongs to the ***Szegő class*** if the logarithm of the absolutely continuous part of the measure is integrable, i.e.

$$\int_0^{2\pi} \log(f(\theta))\frac{d\theta}{2\pi} > -\infty,$$

and this last condition is equivalent to

$$\sum_{n=0}^{\infty} |\Phi_n(0)|^2 < \infty.$$

There is a very useful function associated to σ, which is the so called ***Christoffel function***. One can obtain information about σ from its corresponding Christoffel functions, and it is often said that from the Christoffel functions one "recovers the measure". First, we introduce the positive real numbers

$$\lambda_n(\zeta) = \min\left\{\int_0^{2\pi} |\pi(e^{i\theta})|^2 d\sigma(\theta), \quad \deg \pi \leqslant n, \quad \pi(\zeta) = 1\right\},$$

as a solution of an extremal problem. For a fixed $\zeta \in \mathbb{C}$, λ_n is a positive function which decreases with n. We then call ***Christoffel function*** to the limit

$$\lambda_\infty(\zeta) = \lim_{n\to\infty} \lambda_n(\zeta).$$

The information that can be obtained from the Christoffel function is collected in the following result

Theorem 4 ([28]).

(i) *If* $|\zeta| > 1, \lambda_\infty(\zeta) = 0$.
(ii) *If* $|\zeta| = 1, \lambda_\infty(\zeta) = \sigma(\{\zeta\})$.
(iii) *If* $\sum_{n=0}^{\infty} |\Phi_n(0)|^2 = \infty$, *then* $\lambda_\infty(\zeta) = 0$ *for every* ζ *with* $|\zeta| < 1$.
(iv) *If* $\sum_{n=0}^{\infty} |\Phi_n(0)|^2 < \infty$, *then* $\lambda_\infty(\zeta) > 0$ *for every* ζ *with* $|\zeta| < 1$.

Furthermore,

$$\lambda_\infty(0) = \prod_{n=1}^{\infty}(1 - |\Phi_n(0)|^2).$$

For more information about analytic and spectral theory of OPUC the reader can check [28, 29] as well as the recent survey paper [18].

Following the analogies with the case of the real line, we introduce an analog to the Stieltjes function in $\mathbb{R}$, which here is called the ***Carathéodory function***, and it has certain interesting properties around the origin. From a sequence of moments $\{c_k\}_{k\in\mathbb{Z}}$, it is formally defined as

$$F(z) = c_0 + 2\sum_{n=1}^{\infty} c_{-n} z^n.$$

If $\mathcal{L}$ is a positive definite linear functional, then F is analytic on $\mathbb{D} = \{z \in \mathbb{C} : |z| < 1\}$ and $\Re e(F(z)) > 0$ on $\mathbb{D}$. In this case, F has an integral representation in terms of the Riesz-Herglotz formula

$$F(z) = \int_{\mathbb{T}} \frac{w + z}{w - z} d\sigma(w).$$

Our next step is to study the matrix associated with the multiplication operator by z, that is denoted M_z. Its matrix representation in terms of the basis $\{\varphi_n\}_{n\geq 0}$ is

$$z\varphi(z) = \mathbf{H}_\varphi \varphi(z),$$

where $\varphi(z) = [\varphi_0(z), \varphi_1(z), \ldots, \varphi_n(z), \ldots]^T$ and $\mathbf{H}_\varphi$ is a lower Hessenberg matrix, whose entries $h_{n,j}$ depend on the Verblunsky parameters

$$h_{n,j} = \begin{cases} \dfrac{\kappa_n}{\kappa_{n+1}} & \text{if } j = n+1, \\[2ex] -\dfrac{\kappa_j}{\kappa_n}\Phi_{n+1}(0)\overline{\Phi_j(0)} & \text{if } j \leqslant n, \\[2ex] 0 & \text{if } j > n+1, \end{cases}$$

where κ_n is the leading coefficient of $\varphi_n(z)$. This representation of the operator M_z is called in the literature [28] ***GGT Representation***.

The matrix $\mathbf{H}_\varphi$ satisfies the following properties:

(i) $\mathbf{H}_\varphi \mathbf{H}_\varphi^* = \mathbf{I}$,

(ii) $\mathbf{H}_{\varphi}^{*}\mathbf{H}_{\varphi} = \mathbf{I} - \lambda_{\infty}(0)\varphi(0)\varphi(0)^{*}$, where $\mathbf{I}$ is the identity matrix.

Here, $\varphi(0) = (\varphi_0(0), \varphi_1(0), \varphi_2(0), \ldots)^{*}$.

Notice that in *(ii)*, a rank one matrix appears that prevents $\mathbf{H}_{\varphi}$ from being a unitary matrix, a feature that would be desirable. Due to this fact, this type of matrices are called "almost unitary". It is evident that if for this problem the Christoffel function checks $\lambda_{\infty}(0) = 0$, $\mathbf{H}_{\varphi}$ would be unitary. Indeed, $\lambda_{\infty}(0) = 0$ if and only if the measure σ does not belong to the Szegő class, and, therefore, in such a situation $\mathbf{H}_{\varphi}$ is unitary. The fact that σ belongs to the Szegő class also implies that the OPUC with respect to σ constitutes a complete orthogonal system or, equivalently, $\mathbb{P}$ is dense in the space L_{σ}^{2}.

Next we are going to study perturbations of these Hessenberg matrices.

3.2 Spectral Transformations

Let consider the linear space of complex sequences $\mathbf{D}^{\infty} = \{\{\alpha_n\}_{n\geq 1}, |\alpha_n| < 1\}$, and $\Theta : \mathbf{D}^{\infty} \to \mathbf{D}^{\infty}$ a 1-1 mapping. According to the Verblunsky theorem, Θ induces an application $\Upsilon : \mathcal{P} \to \mathcal{P}$, in the set $\mathcal{P}$ of probability measures supported on the unit circle such that the following diagram

$$\begin{array}{ccc} \mathcal{P} & \xrightarrow{\Upsilon} & \mathcal{P} \\ \mathcal{S}\downarrow & & \downarrow\mathcal{S} \\ \mathbf{D}^{\infty} & \xrightarrow{\Theta} & \mathbf{D}^{\infty} \end{array}$$

is commutative, that is, $\Theta \circ \mathcal{S} = \mathcal{S} \circ \Upsilon$. The application Υ is called a ***spectral transformation***. Likewise, each application $\mathcal{F}$ on the family of Carathéodory-functions F, such that $\mathcal{F}(F)$ is also a Carathéodory-function, induces a spectral transformation Υ so that the diagram is commutative. We will also call such applications spectral transformations. In what follows, we will show some spectral transformations on $\mathcal{P}$, from certain perturbations on the Verblunsky coefficients $\{\Phi_n(0)\}_{n\geq 1}$ associated with a measure of probability σ, or from spectral transformations on a C-function.

3.2.1 Canonical Linear Spectral Transformations

Again, we find a clear parallelism with the case of the real line. Canonical linear transformations in $\mathbb{T}$ are defined as follows see [11]

1. Christoffel transformation ($\mathcal{F}_C$)

$$d\tilde{\sigma} = |z - \alpha|^2 d\sigma, \quad \alpha \in \mathbb{C} \setminus \{0\}.$$

2. Uvarov transformation ($\mathcal{F}_U$)

$$d\tilde{\sigma} = d\sigma + M_c\delta(z - \alpha) + \overline{M}_c\delta(z - \bar{\alpha}^{-1}), \quad \alpha \in \mathbb{C}\setminus\{0\}, \quad M_c \in \mathbb{C}.$$

3. Geronimus transformation ($\mathcal{F}_G$)

$$d\tilde{\sigma} = \frac{d\sigma}{|z-\alpha|^2} + M_c\delta(z-\alpha) + \overline{M}_c\delta(z-\bar{\alpha}^{-1}), \quad \alpha \in \mathbb{C}\setminus\{0\}, \quad M_c \in \mathbb{C}.$$

Note that for the Uvarov and Geronimus transformations, two different cases must be distinguished. One case when the mass is on the unit circle, and the other one when the mass is a point outside the unit circle. In the first one we will have $\alpha = (\bar{\alpha})^{-1}$ and then the perturbation is real. In the second one we have two complex conjugated masses located at inverse points with respect to the unit circle. In both cases, the Toeplitz structure of the Gram matrix is preserved. These three canonical perturbations satisfy

$$\begin{aligned} &\mathcal{F}_C \circ \mathcal{F}_G = I \quad \text{(Identity transformation)}, \\ &\mathcal{F}_G \circ \mathcal{F}_C = \mathcal{F}_U. \end{aligned}$$

3.2.2 Rational Spectral Transformations

The above transformations appear in the framework of perturbations of probability measures supported on $\mathbb{T}$, but one can deal with perturbations of Verblunsky coefficients as follows.

1. Second kind polynomials ($\mathcal{F}_\Omega$) associated with the Carathéodory function $\mathcal{F}_\Omega = F^{-1}$, which arise changing the sign the Verblunsky coefficients

$$\Omega_n(0) = -\Phi_n(0), \quad n \geq 1.$$

2. Aleksandrov transformation ($\mathcal{F}^\lambda$), corresponding to a multiplication of the Verblunsky coefficients by a complex number of modulus one

$$\Phi_n^\lambda(0) = \lambda\Phi_n(0), \quad |\lambda| = 1, \quad n \geq 1.$$

3. Associated polynomials of order N ($\mathcal{F}^{(N)}$)

$$\Phi_n^{(N)}(0) = \Phi_{n+N}(0), \quad n \geq 1.$$

4. Anti-associated polynomials of order N ($\mathcal{F}^{(-N)}$)

$$\{\Phi_n^{(-N)}(0)\}_{n\geq 1} = \{v_i\}_{i=1}^N \bigcup \{\Phi_n(0)\}_{n\geq 1}, \quad |v_i| < 1.$$

They can be constructed by modifying the Verblunsky coefficients in the recurrence relation $\Phi_{n+1}(z) = z\Phi_n(z) + \Phi_{n+1}(0)\Phi_n^*(z)$, with $\Phi_n^*(z) = z^n\overline{\Phi}_n(z^{-1})$.

3.2.3 Carathéodory functions and spectral transformations

The study of spectral transformations for OPRL presented in Sect. 2 can be reproduced in the OPUC framework. Indeed, you can introduce linear spectral transformations

$$\widetilde{F}(z) = \frac{A(z)F(z) + B(z)}{D(z)},$$

as well as rational spectral transformations

$$\widetilde{F}(z) = \frac{A(z)F(z) + B(z)}{C(z)F(z) + D(z)}, \quad AD - BC \neq 0.$$

As an illustrative example in the matrix perspective, let us see how the Hessenberg matrix is affected through the Christoffel transformation. Let $\mathbf{H}_{\varphi}$ be the Hessenberg matrix associated with a non-trivial probability measure σ supported on $\mathbb{T}$. If $d\tilde{\sigma} = |z - \alpha|^2 \, d\sigma$ is a Christoffel transformation of the measure σ, and $\psi(z) = [\psi_0(z), \psi_1(z), \ldots]^T$ is the corresponding orthonormal polynomial vector, we have that $(z-\alpha)\psi(z) = \mathbf{M}_c\varphi(z)$. Taking into account the matrix of the change of bases $\varphi(z) = \mathbf{L}\psi(z)$, where $\mathbf{L}$ is a lower triangular matrix, we have the factorization

$$\begin{aligned} \mathbf{H}_{\varphi} - \alpha\mathbf{I} &= \mathbf{L}\mathbf{M}_c \\ \mathbf{H}_{\psi} - \alpha\mathbf{I} &= \mathbf{M}_c\mathbf{L}, \end{aligned}$$

with

$$\begin{aligned} \mathbf{M}_c\mathbf{M}_c^* &= \mathbf{I} \\ \mathbf{M}_c^*\mathbf{M}_c &= \mathbf{I} - \lambda_{\infty}(\alpha)\varphi(\alpha)\,[\varphi(\alpha)]^* . \end{aligned}$$

Again, we get that $\mathbf{M}_c$ is an almost-unitary matrix. Thus, we still need that the OPUC with respect to σ be dense in L^2_{σ} to have a suitable representation.

3.3 CMV Representation

There is another representation of the z multiplication operator M_z, that is called the ***CMV representation*** (see [10]). Its main advantage is that it removes the restriction for the measure (to be outside the Szegő class) in order the matrix is unitary. Instead of a polynomial basis, an orthogonal basis in the linear space of Laurent polynomials is required. In this way, the multiplication operator can be represented by a five-diagonal matrix called the CMV matrix, which also admits a factorization as a product of two tridiagonal matrices.

The key idea is to organize the basis of the Laurent polynomial space in a more convenient way, say $\{1, z, z^{-1}, z^2, z^{-2}, \ldots\}$, and then apply the Gram-Schmidt orthogonalization process to the elements of this basis. Thus, one has a basis $\{x_n\}_{n\geq 0}$ given in terms of the OPUC as follows

$$\begin{aligned} x_{2n}(z) &= z^{-n}\varphi^*_{2n}(z), \\ x_{2n+1}(z) &= z^{-n}\varphi_{2n+1}(z). \end{aligned}$$

The representation of the multiplication operator with respect to the above orthogonal basis of Laurent polynomials is then the aforementioned five-diagonal matrix

$$C = \mathcal{L}\mathcal{M},$$

where $\mathcal{L}$ $=diag(\theta_0, \theta_2, \theta_4, \ldots)$, and $\mathcal{M} = diag(1, \theta_1, \theta_3, \ldots)$, with

$$\theta_j = \begin{bmatrix} -\Phi_{j+1}(0) & \rho_{j+1} \\ \rho_{j+1} & \overline{\Phi_{j+1}(0)} \end{bmatrix}, \quad \rho_{j+1} = (1 - |\Phi_{j+1}(0)|^2)^{\frac{1}{2}}. \tag{16}$$

The above matrix is always unitary.

Linear spectral transformations of nontrivial probability measures supported on the unit circle and the corresponding CMV representations have been studied in [11].

Problem 1. *To find the CMV representation of rational spectral perturbations.*

Notice that for the Aleksandrov transformation, the following result is very well known (see [28])

Theorem 5. *Let* $|\lambda| = 1$ *and* $\widetilde{\Phi}_n(0) = \lambda\Phi_n(0)$, $n = 1, 2, \ldots$ *If* $U(\lambda) = diag(\overline{\gamma}, \gamma, \overline{\gamma}, \gamma, \ldots)$, $\gamma^2 = \lambda$, *then* $\widetilde{C} - U(\lambda)CU(\lambda)^{-1}$ *is a rank one operator.*

A survey paper concerning the links between matrix analysis and OPRL (resp. OPUC) is [16].

3.4 Integrable Systems

We now address the issue of integrable systems. In 1975 M. J. Ablowitz and J. F. Ladik (see [1]) consider a spatial discretization of the modified KdV equation

$$u_t = 6u^2 u_x - u_{xxx}$$

appearing the so-called ***Schur flows***. The study of integrable systems in the unit circle has not attracted the interest of researchers in comparison with the case of the real line. Indeed, we can date its development from the end of the last century (see [2,14,23,25]).

3.4.1 Schur Finite Flow

In 1999, L. Faybusovich and M. Gekhtman [14] studied real and finite Schur flows, obtaining the Lax pairs for the GGT Representation. Let us see then how the finite Schur flow is solved, and we will observe that the obtained result is linked to the entries of the corresponding CMV matrix. Thus CMV representation is sometimes said to be motivated by the finite Schur flow solution.

Let $H \in \mathbb{C}^{(n,n)}$ be a upper unitary Hessenberg matrix with non-negative subdiagonal entries. Then it be factorized as

$$H = G_1(\alpha_1)G_2(\alpha_2)\cdots G_{n-1}(\alpha_{n-1})\widetilde{G}_n(\alpha_n), \tag{17}$$

where $|\alpha_j| \leq 1$, $j = 1, 2, \ldots, n-1$, $|\alpha_n| = 1$. $G_j(\alpha)$ are block matrices of the form

$$G_j(\alpha) = \begin{bmatrix} I_{j-1} & 0 & 0 \\ 0 & \Theta(\alpha) & 0 \\ 0 & 0 & I_{n-j-1} \end{bmatrix}.$$

Here $\Theta(\alpha)$ is the same as in (16) when you deal with the CMV representation, i.e.,

$$\Theta(\alpha) = \begin{bmatrix} -\alpha & \rho \\ \rho & \overline{\alpha} \end{bmatrix}, \quad \text{with } \rho = \left(1 - |\alpha|^2\right)^{1/2}.$$

The last factor in (17) is a diagonal matrix of the form

$$\widetilde{G}_n(\alpha) = diag(1, 1, \ldots, 1, -\alpha).$$

The important point to keep in mind now is that H is unitary if and only if $|\alpha| = 1$. Our next step is to study the Lax pair corresponding to this matrix in such a way you get

$$\begin{aligned} \dot{H}(t) &= [H, S(H)], \\ H(0) &= H_0 \in \mathbb{C}^{(n,n)}, \end{aligned}$$

with $S(H)$ the unique anti-hermitian matrix such that $H-S(H)$ is upper triangular with real entries in the main diagonal. Indeed,

Proposition 6. *The matrix $H(t)$ satisfies*

$$H(t) = Q^*(t)H_0Q(t),$$

where $Q(t)$ is a unitary matrix such that

$$\exp H_0 t = Q(t)R(t)$$

and $R(t)$ is an upper triangular matrix with positive entries in the main diagonal. If H_0 is normal with eigenvalues $\lambda_1, \lambda_2, \ldots, \lambda_n$, then

$$H(t) = U(t)\Lambda U^*(t),$$

where $U(t)$ is unitary, and $\Lambda = diag(\lambda_1, \lambda_2, \ldots, \lambda_n)$.

Theorem 6. *If H_0 is a unitary upper Hessenberg matrix, with positive subdiagonal entries, then $H(t)$ is a unitary upper Hessenberg matrix with subdiagonal positive entries. Moreover, if the entries are real, then*

$$\dot{\alpha}_k(t) = (1 - \alpha_k^2(t))(\alpha_{k+1}(t) - \alpha_{k-1}(t))$$

with the initial conditions

$$\begin{aligned} \alpha_0 &\equiv 1, \\ \alpha_n &\equiv \alpha_n(0). \end{aligned}$$

3.4.2 Semi-infinite Schur Flow

Let us assume that $\{\alpha_n(t)\}_{n\geq 1}$ is a sequence of Verblunsky parameters. According to Theorem 3 let $\sigma(z, t)$ be the corresponding probability measure supported on the unit circle, where z is on the unit circle and t is a time parameter. Thus

Theorem 7 ([17])**.** *Given the sequence of complex functions $\{\alpha_n(t)\}_{n\geq 1}$, with $|\alpha_n(t)| < 1$, for $t \geq 0$, and $\alpha_{-1} \equiv -1$, the following statements are equivalent.*

(i) The sequence $\{\alpha_n(t)\}_{n\geq 1}$ *satisfies the* ***Schur flow equations***

$$\dot{\alpha}_n(t) = (1 - |\alpha_k(t)|^2)(\alpha_{n+1}(t) - \alpha_{n-1}(t))$$

with $n \in \mathbb{Z}^+$, $t > 0$.

(ii) The CMV matrix associated with the sequence $\{\alpha_n(t)\}_{n\geq 1}$ *satisfies*

$$C(t) = [A, C],$$

where

$$A = \begin{bmatrix} \Re e(\overline{\alpha}_0) & \rho_0 \overline{A}_0 & \rho_0\rho_1 & 0 & 0 & 0 \\ & -\Re e(\overline{\alpha_1}\alpha_0) & \rho_0 A_1 & \rho_1\rho_2 & 0 & 0 \\ & 0 & -\Re e(\overline{\alpha_2}\alpha_1) & \rho_2\overline{A}_2 & \rho_2\rho_3 & 0 \\ & 0 & 0 & \ddots & \ddots & \ddots \end{bmatrix}$$

with $A_n = \alpha_{n+1}(t) - \alpha_{n-1}(t)$, $\rho_n(t) = (1 - |\alpha_n(t)|^2)^{1/2}$.

(iii) The probability measure $\sigma(z,t)$ *supported on the unit circle, such that* $\{\alpha_n(t)\}_{n\geq 1}$ *is the corresponding sequence of Verblunsky parameters, satisfies*

$$d\sigma(z,t) = C(t) \exp t(z + z^{-1}) d\sigma(z,0),$$

where $C(t)$ *is a normalization coefficient such that* $\sigma(\mathbb{T}, t) = 1$.

A nice survey concerning sequences of OPRL and OPUC polynomials, Toda lattices and Painlevé equations is [32].

Problem 2. *To study the flows associated with the following perturbations*

$$d\sigma(z,t) = |z - t|^2 \, d\sigma(z,0),$$
$$d\sigma(z,t) = (1 - M(t))d\sigma(z,0) + M(t)\delta(z - e^{i\theta_0}), \text{ with } M(0) = 0$$

which correspond to Christoffel and Uvarov transformations of the initial measure $d\sigma(z,0)$ *respectively.*

Problem 3. *According to the results in [30] about zeros of OPUC associated with Schur flows, it is a remarkable question to analyze the dynamics of zeros of OPUC corresponding to the perturbations considered in the above Problem 2.*

Acknowledgements. We thank the careful revision by the referees. Their suggestions and comments have improved the presentation of the manuscript.

The work of E. J. Huertas has been supported by Dirección General de Investigación e Innovación, Consejería de Educación e Investigación of the Comunidad de Madrid (Spain) and Universidad de Alcalá, under grant CM/JIN/2021-014, *Proyectos de I+D para Jóvenes Investigadores de la Universidad de Alcalá 2021*, and the Ministerio de Ciencia e Innovación-Agencia Estatal de Investigación MCIN/AEI/10.13039/501100011033 and the European Union "NextGenerationEU"/PRTR, under grant TED2021-129813A-I00.

The work of F. Marcellán has been supported by FEDER/Ministerio de Ciencia e Innovación-Agencia Estatal de Investigación of Spain, grant PID2021-122154NB-I00, and the Madrid Government (Comunidad de Madrid-Spain) under the Multiannual Agreement with UC3M in the line of Excellence of University Professors, grant EPUC3M23 in the context of the V PRICIT (Regional Program of Research and Technological Innovation).

References

1. Ablowitz, M.J., Ladik, J.F.: Nonlinear differential-difference equations and Fourier analysis. J. Math. Phys. **17**, 1011–1018 (1976)
2. Ammar, G.S., Gragg, W.B.: Schur flows for orthogonal Hessenberg matrices. In: Hamiltonian and Gradient Flows, Algorithms and Control. Fields Institute Communications, vol. 3, pp. 27–34. American Mathematical Society, Providence (1994)
3. Aptekarev, A.I., Branquinho, A., Marcellán, F.: Toda-type differential equations for the recurrence coefficients of orthogonal polynomials and Freud transformation. J. Comput. Appl. Math. **78**(1), 139–160 (1997)
4. Rolanía, D.B., Branquinho, A., Moreno, A.F.: On the full Kostant-Toda system and the discrete Korteweg-de Vries equations. J. Math. Anal. Appl. **401**(2), 811–820 (2013)
5. Rolanía, D.B.: On the Darboux transform and the solutions of some integrable systems. Rev. R. Acad. Cienc. Exactas Fís. Nat. Ser. A Mat. RACSAM **113**(2), 1359–1378 (2019)
6. Berezanski, Y.M.: The integration of semi-infinite Toda chain by means of inverse spectral problem. Rep. Math. Phys. **24**, 21–47 (1986)
7. Bueno, M.I., Marcellán, F.: Darboux transformation and perturbation of linear functionals. Linear Algebra Appl. **384**, 215–242 (2004)
8. Brezinski, C.: A brief introduction to integrable systems. Int. J. Comput. Sci. Math. **1**(1), 98–106 (2007)
9. Buhmann, M.D., Iserles, A.: On orthogonal polynomials transformed by the QR algorithm. J. Comput. Appl. Math. **43**(1–2), 117–134 (1992)
10. Cantero, M.J., Moral, L., Velázquez, L.: Five-diagonal matrices and zeros of orthogonal polynomials on the unit circle. Linear Algebra Appl. **362**, 29–56 (2003)
11. Cantero, M.J., Marcellán, F., Moral, L., Velázquez, L.: Darboux transformations for CMV matrices. Adv. Math. **298**, 122–206 (2016)
12. Chihara, T.S.: An Introduction to Orthogonal Polynomials. Gordon and Breach Science Publishers, New York (1978)
13. Demmel, J.W.: Applied Linear Algebra. Cambridge University Press, Cambridge (1997)
14. Faybusovich, L., Gekhtman, M.: On Schur flows. J. Phys. A Math. Gen. **32**, 4671–4680 (1999)
15. García-Ardila, J.C., Marcellán, F., Marriaga, M.E.: Orthogonal Polynomials and Linear Functionals-An Algebraic Approach and Applications EMS Series of Lectures in Mathematics. EMS Press, Berlin (2021)
16. Garza, L.E., Marcellán, F.: Orthogonal polynomials and perturbations on measures supported on the real line and on the unit circle. A matrix perspective. Expo. Math. **34**(3), 287–326 (2016)
17. Golinskii, L.B.: Schur flows and orthogonal polynomials on the unit circle. Mat. Sb. **197** (8), 41–62 (2006) (Russian); translation in Sb. Math. **197**(7–8), 1145–1165 (2006)
18. Golinskii, L.B.: Orthogonal polynomials on the unit circle. Encyclopedia of special functions: the Askey-Bateman project. Vol. 1. Univariate orthogonal polynomials, pp. 199–241, Cambridge University Press, Cambridge (2020)
19. Hermoso, C., Huertas, E.J., Lastra, A., Marcellán, F.: Higher-order recurrence relations, Sobolev-type inner products and matrix factorizations. Numer. Algorithms **92**, 665–692 (2023)

20. Kasman, A.: Orthogonal polynomials and the finite Toda lattice. J. Math. Phys. **38**, 247–254 (1997)
21. Mañas, M.: Revisiting biorthogonal polynomials: an LU factorization discussion. In: Marcellán, F., Huertas, E.J. (eds.) Orthogonal Polynomials: Current Trends and Applications. SSSS, vol. 22, pp. 273–307. Springer, Cham (2021). https://doi.org/10.1007/978-3-030-56190-1_10
22. Moser, J.K.: Finitely many mass points on the line under the influence of an exponential potential-an integrable system. In: Moser, J. (ed.) Dynamical Systems, Theory and Applications. Lecture Notes in Physics, vol. 38, pp. 467–497. Springer, Heidelberg (1975). https://doi.org/10.1007/3-540-07171-7_12
23. Mukaihira, A., Nakamura, Y.: Schur flow for orthogonal polynomials on the unit circle and its integrable discretization. J. Comput. Appl. Math. **139**(1), 75–94 (2002)
24. Nakamura, Y., Sekido, H.: An application of the discrete-time Toda lattice to the progressive algorithm by Lanczos and related problems. J. Phys. A:Math. Theor. **51**(17), 174001 (2018)
25. Nenciu, I.: Lax pairs for the Ablowitz-Ladik system via orthogonal polynomials on the unit circle. Int. Math. Res. Not. **2005**(11), 647–686 (2005)
26. Nikishin, E.M., Sorokin, V.N.: Rational approximations and orthogonality. Translations of Mathematical Monographs, vol. 92. American Mathematical Society, Providence (1991)
27. Peherstorfer, F., Spiridonov, V.P., Zhedanov, A.: The Toda chain, the Stieltjes function, and orthogonal polynomials. Teoret. Mat. Fiz. **151**(1), 81–108 (2007) (Russian); translation in Theoret. and Math. Phys. **151**(1), 505–528 (2007)
28. Simon, B.: Orthogonal polynomials on the unit circle. Part 1. Classical theory. American Mathematical Society Colloquium Publications, vol. 54, Part 1. American Mathematical Society, Providence (2005)
29. Simon, B.: Orthogonal polynomials on the unit circle. Part 2. Spectral theory. American Mathematical Society Colloquium Publications, vol. 54, Part 2. American Mathematical Society, Providence (2005)
30. Simon, B.: Zeros of OPUC and long time asymptotics of Schur and related flows. Inverse Probl. Imaging **1**(1), 189–215 (2007)
31. Spiridonov, V., Vinet, L., Zhedanov, A.: Spectral transformations, self-similar reductions and orthogonal polynomials. J. Phys. A **30**(21), 7621–7637 (1997)
32. Van Assche, W.: Orthogonal polynomials, Toda lattices and Painlevé equations. Phys. D **434**, Paper No. 133214, 9 pp (2022)
33. Vinet, L., Zhedanov, A.: An integrable system connected with the Uvarov-Chihara problem for orthogonal polynomials. J. Phys. A **31**(47), 9579–9591 (1998)
34. Zhedanov, A.: Rational spectral transformations and orthogonal polynomials. J. Comput. Appl. Math. **85**, 67–83 (1997)

Author Index

H. Herrero et al. (Eds.): RSME 2022, RSME 13, p. 153, 2024.
https://doi.org/10.1007/978-3-031-62025-6

The manufacturer's authorised representative in the EU is Springer Nature Customer Service Centre GmbH, Europaplatz 3, 69115 Heidelberg, Germany. If you have any concerns regarding our products, please contact ProductSafety@springernature.com

Printed and bound by CPI Group (UK) Ltd, Croydon, CR0 4YY
28/09/2026
02224070-0016